The Dependability Revolution: Mathematical Tools for Building Secure, Reliable Systems

Sanjay

Contents

Introduction

1

Dependability as a human characteristic has always been appreciated. In the middle of the last century, the development of a whole theory with respect to dependability in an engineering context became a distinct research discipline, driven by the objective of building a technical system to be able to survive the moon mission. Since then, dependability theory has been providing mathematical tools to evaluate life cycles and prevent or reduce failures of any technical system, e. g., electronic equipment, vehicles, machinery, and industrial plants. Reliability is only one of the aspects of dependability, which includes also security, integrity, safety, maintainability, and availability. Targeting wireless reliability and even "ultra reliability" has gained momentum with the research on the fifth generation (5G) of mobile communications systems and beyond, which also lends impetus to this book. This chapter elaborates this motivation and presents the contributions of the book as an outline.

1.1 Motivation

A main objective of 5G wireless communications systems is the support of diverse applications in a flexible and reliable way. Besides enhanced Mobile Broadband (eMBB) and massive Machine-Type Communications (mMTC), ultra-reliable low-latency communications (URLLC) is the third pillar of 5G, as agreed by academia, industry, and standardization groups [ITU15; 3GP17b; Pop+18a].

The primary goals regarding eMBB are capacity and data rate enhancements for multimedia and entertainment applications. Thus, eMBB can be considered as the extension to the 4G broadband service, which is used by humans. In contrast, mMTC and URLLC fall into the category of wireless machine-to-machine (M2M) communication. The domain of mMTC is expected to support a massive number of Internet of Things (IoT) devices, e. g., wireless sensors and actuators for smart city, smart logistics, and smart home applications [Sha+15]. Since these devices are only sporadically active and send small data payloads, wide coverage and energy efficiency is important. By ensuring a latency in the (sub-)

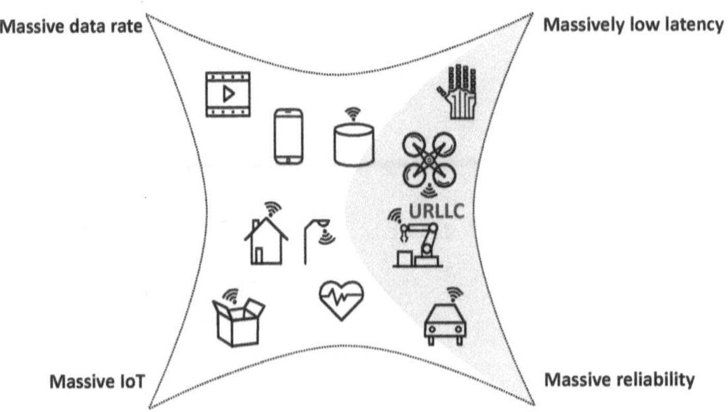

Fig. 1.1.: Dimensions of challenges for 5G and beyond.

millisecond range together with ultra-high reliability in terms of packet loss ratio (PLR) of 10^{-5} up to 10^{-9}, URLLC is expected to enable unprecedented use cases, e.g., wireless factory automation, self-driving cars, and real-time remote control, paving the way towards the Tactile Internet [Sim+16; Fet14; Sch+17]. URLLC is especially challenging, because it depends on the simultaneous fulfillment of strict requirements in terms of latency and reliability. These requirements might be contradictory, e. g., reliability can be achieved through retransmissions, but this increases the latency. However, there are several use cases which do not require this combination, e. g., online games depend on low latency but no severe damage will follow if a pixel is wrong. On the other hand, autonomous cars may perform precise maneuvers at slower speeds tolerating higher communication latency. Consequently, the various challenges for 5G and beyond can be arranged in a space spanned by four dimensions, as proposed in Fig.1.1, which contrasts with the well-known triangle of 5G applications initially published by [ITU15]. Different terminology evolved, implying that the requirements for latency and reliability should first be considered separately, e. g., ultra-reliable communication (URC) [Pop14; BDP18], ultra-reliable machine-type communication (uMTC) [MET16], or critical machine-type communication (cMTC) [Sch20].

In 2018, the 3rd Generation Partnership Project (3GPP) published standardization specifications regarding the first phase of 5G (Release 15), which includes, e. g., the New Radio (NR) air interface, the new radio network architecture called next generation radio access network (NG-RAN), the new core network architecture called 5G core (5GC), network slicing and edge computing [Roh20]. The work accomplished in Release 15 focused on the eMBB usage scenario, URLLC functionalities are planned to be incorporated in the second phase of 5G (Release 16), to be published in June 2020. However, URLLC will continue to constitute an important topic of research even beyond 5G, i. e., in the next generation of mobile communications systems, 6G, which is expected to launch on a commercial basis by 2030. The first 6G white papers outline the vision of "enhanced" or "extreme" URLLC [NTT20; Jun+19]. Although the terms become more dramatic, many of the key performance indicators (KPIs) used for developing 5G technologies are envisioned to remain valid. This holds true, e.g., for the use cases of wireless industrial control, which are identified as the most demanding in terms of latency and reliability, because no more than one lost packet in a billion transmitted packets with a $0.1\,\text{ms}$ latency is permitted [Jun+19; Sch20].

Obviously, these requirements are driven by the objective to utilize wireless links as replacement of wired field bus connections, which are claimed to provide average PLRs on application level in the range of 10^{-9} [Fro+14; Sch+20a]. Unfortunately, expressing reliability as the probability of receiving the correct packet before its deadline, which is common in the communications sector, can only be translated into the time domain to a limited extent: A PLR of 10^{-9} is equivalent to a sum of only $31.5\,\text{ms}$ outage time during one year of non-stop operation. Since this outage duration is the accumulated value, it is not possible to determine more precise conditions on tolerated outage bursts, e.g., it is not covered, if each month allows $2.6\,\text{ms}$ of outage or if an outage burst of $31.5\,\text{ms}$ has to be be followed by a whole year without any lost packet. Furthermore, the assumption of non-stop operation does not reflect reality adequately, e.g., in industrial control or automated driving. Instead, intervals with different reliability requirements, which follow each other, are of interest, as illustrated by the following examples. High accuracy pick and place operations performed by a wirelessly controlled robot may include waiting times for new work pieces, which can accept a higher number of packet losses. On the other hand, a self-driving car might need increased connectivity reliability without interruptions during critical maneuvers such as crossing an intersection. Since the common interpretation of reliability in wireless communications fails in expressing these various aspects related to the dependency on the time dimension, the introduction of new KPIs can be beneficial in order to refine the discussion on URLLC. In this regard, dependability theory is worth receiving more attention, because it provides a framework of well-accepted terminology, KPIs, mathematical concepts, and methods to model and analyze life cycles and failures of technical systems. Especially the fact that it is often essential to guarantee a performance level necessary for successfully completing a certain mission during a time interval resembles the objective to survive the first moon missions, which was an incentive to constitute dependability theory. But even though the concept of dependability combines the consideration of different aspects including, e.g., reliability, availability, and time to failure, its tool set has so far hardly been exploited in wireless communications engineering.

This book contributes to fill this gap by focusing on dependable wireless communication through multi-connectivity. Multi-connectivity summarizes approaches that establish multiple communication links simultaneously and is considered to be a promising key approach in order to increase wireless reliability [Öhm17; BDP18; Wol+19]. It corresponds to the general dependability concept of redundancy, i. e., employing one or more reserve components. However, especially in the case of multiple users, who compete for limited wireless resources, adding links may not always be beneficial. This book, thus, also addresses the question of how to achieve a stable resource allocation for dependable wireless communication in multi-user, multi-cellular systems.

1.2 Contribution

The overall objective of this book is to provide insights on how to leverage dependability theory aiming for dependable wireless communication through multi-connectivity from the perspective of a single user as well as for multi-user systems. The following contributions are presented in this book:

- Chapter 2 lays the foundation of the book by presenting an overview on the state of the art and relevant basics related to dependability and multi-connectivity.

- In Chapter 3, Markov models are developed which describe a wireless multi-connectivity approach and selected causes of failures in terms of dependability theory and, thus, form the basis for subsequent evaluations. By interpreting the wireless channel as a repairable system component, corresponding to the Gilbert-Elliot model, continuous-time Markov chains (CTMCs) are derived. They model whether both the individual components and the overall multi-connectivity system are in failure or can function correctly. The relevant causes of failure in wireless communications which this chapter focuses on are co-channel interference and small-scale fading according to Rayleigh fading and Rice fading, respectively.

- Chapter 4 proposes and evaluates appropriate KPIs based on dependability theory with respect to the Markov models, introduced in Chapter 3. Potential dependability gains due to multi-connectivity are discussed for selection-combined channels. Especially, the benefit of time-related and mission-related KPIs is demonstrated. Besides the discussion of dependability metrics characterizing average performance, the probability distributions of up- and downtimes are evaluated, which enable concrete performance guarantees. By proposing the metric "mission reliability", it is demonstrated how the success probability of a continuous mission, which does not allow a failure, depends on its mission duration. This KPI is also extended to robust applications, denoting systems which can tolerate short outage times. Thus, Chapter 4 leverages dependability theory in order to propel the realization of wireless URLLC through multi-connectivity by refining the discussion on appropriate KPIs.

- In Chapter 5, the novel dynamic connectivity concept reaction diversity is proposed and evaluated. In contrast to static connectivity approaches, it dynamically reacts to fades by selecting a different channel. Therefore reaction diversity avoids wasting resources that are not needed for a smooth operation, as it is the case with static multi-connectivity. By means of numerical simulations considering Rayleigh fading as the cause of failure, the outage probability as well as distributions of uptime and downtime with respect to robust applications utilizing reaction diversity are evaluated and compared to state-of-the-art connectivity approaches.

- Complementing the previous chapters, Chapter 6 focuses on dependable wireless communication through multi-connectivity in multi-user, multi-cellular systems, which pose the challenges of interference and the competition for limited re-sources. Analytic comparisons of different connectivity approaches are developed, discussing why adding links may not always be optimal in the considered scenario. The resource allocation is modeled based on matching theory aiming for stability, i.e., no pair of users and allocated resources should have an incentive to switch partners. Extensions to the many-to-one stable matching procedure are proposed targeting the requirements of wireless URLLC. The novel resource allocation strategies utilize the optimal connectivity approach for each user, optimize the maximum number of matched resources, and provide a resource reservation mech-anism for users suffering from bad channel conditions. System-level simulations are evaluated and compared to baseline resource allocation strategies in terms of outage probability.

- Finally, Chapter 7 concludes this book and highlights open questions for further research.

State of the Art

2

This book links dependability theory and multi-connectivity. An overview of relevant related literature regarding both areas is provided in this chapter. First, fundamental terminology and concepts of dependability theory are introduced before the state of the art regarding multi-connectivity is presented.

2.1 Fundamentals of Dependability Theory

One of the first contributions to dependability theory is known as Lusser's Law: After World War II, the German engineer Robert Lusser formalized the product law for reliability in series systems, i. e., the reliability of a series of components is equal to the product of the individual reliabilities of the components, assuming statistically independent failures. This implies that systems comprising a large number of components may suffer from a rather low system reliability, even though the individual components have high reliabilities [N R95]. Since the industrial development has been growing and products have been continuing to become more complex, consisting of an increasing number of components, dependability theory gained in importance. Dependability theory was particularly fostered during the "Space Race", i. e., the technical advancement of space travel staged like a competition between the USA and USSR which peaked with the landing of the first humans on the Moon in 1969. Pioneering work in textbooks and the first journal on this topic, the IEEE Transactions on Reliability, appeared in the 1960s [BP65; IEE63]. Since then, the importance and advancement of dependability theory as a tool to analyze the life cycles and failures of technical systems has been reflected in the steadily growing body of literature.

An ongoing controversy concerns the question of which basic term – reliability, dependability, or availability – is the most general concept. In the context of this book, the following systematic categorization is utilized according to [Avi+04]: Dependability theory is the comprehensive framework, involving the main attributes availability, reliability, maintainability, safety, integrity, and security, as illustrated in Fig. 2.1. Thus, dependability theory covers the development of mathematical methods in order to evaluate and demonstrate these aspects of technical components, equipment, and systems [Bir13]. However, only availability and reliability are quantifiable in terms of probabilities for correct service and its continuity, respectively. Thus, an important difference between reliability and availability is that reliability refers to failure-free operation during a time

Fig. 2.1.: Dependability theory attributes.

interval, while availability refers to failure-free operation at a given instant of time [Gro+08]. Reliability is also referred to as "survivor function" because it corresponds to the probability that a considered technical item does not fail, i. e., it survives, during a certain time interval. Availability and reliability coincide only in the special case when the item is not repaired after a failure [HR09].

Important quantities, related to the probabilities availability and reliability, are expressed in terms of time duration. An overview of respective dependability KPIs, which are covered and extended in this book, is presented as follows:

- Probabilities:

    - Availability,

    - Reliability.

- Time duration:

    - Mean uptime (MUT),

    - Mean downtime (MDT),

    - Mean time to first failure (MTTFF),

    - Mean time between failures (MTBF).

These measures are given a mathematically precise definition in Chapter 4. A common assumption is the confinement to situations where it suffices to consider only a functioning state and a failure state for each component as well as the system [HR09]: Following the usual notion, "up" denotes an operating state and "down" refers to a failed state, i. e., in repair if repairs are possible. A common condition for all quantities is that the considered item is operational at the beginning of the observation.

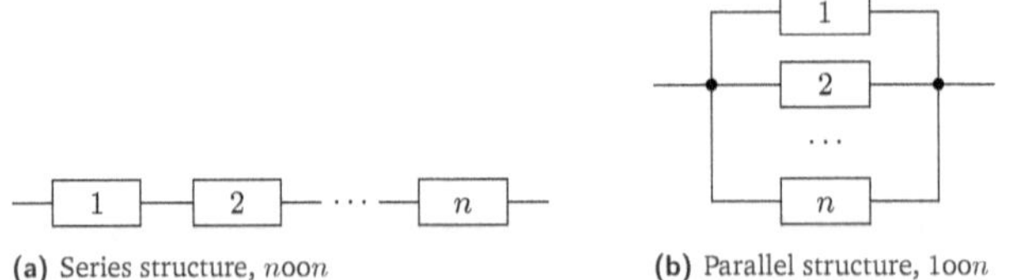

(a) Series structure, $noon$ (b) Parallel structure, $1oon$

Fig. 2.2.: Reliability block diagrams for series and parallel structure as special cases of the $koon$ structure.

Since a technical system is often composed of several components, one key approach to improve availability and reliability of a system is to introduce redundancy, i.e., employ one or more reserve components. A generic notation to express the concept of redundancy is the k-out-of-n ($koon$) structure. It characterizes a system which is functioning if and only if at least k of the n components are operational [KZ03]. This can be illustrated by a reliability block diagram showing the logical connections of components necessary to fulfill the system function [Tri16]: A $noon$ refers to a series structure without redundancy whereas a parallel structure corresponds to $1oon$ realizing full redundancy, as visualized in Fig. 2.2. Further fundamental concepts of dependability theory, which are relevant in the context of this book, comprise structure functions, stochastic failure modeling, repairable systems, and CTMCs. These tools are discussed in the thorough and comprehensive introductions to component and system dependability analysis in [HR09; Bir13; Tri16].

The definitions and concepts of dependability theory can also be applied to communications. Exemplary contributions include the assessment of reliability and performance of information technology systems by means of probabilistic, discrete-state models [STP12], and examination of security aspects of communications systems from a dependability theory perspective [XGT14]. The expected reliability and MTTFF of a wireless network system with imperfect components are analyzed in [CL05]. However, perfect communication links are assumed. Approaches of adopting dependability theory concepts in order to enhance the dependability of wireless networks are presented in [LQ10], where the authors only consider the steady-state probability of channel availability instead of investigating transition probabilities between operational and failed states. Available and unavailable time intervals are modeled based on the channel occupancy status in [BLP17]. However, this availability and reliability analysis is restricted to cognitive radio networks.

Within the communications domain, KPIs such as "availability" and "reliability" are often used interchangeably and not in accordance with with the terminology of traditional dependability theory, even though the International Telecommunication Union (ITU) has adopted the definitions to communications [ITU08]. According to [3GP17c], reliability is considered as the percentage of successful transmissions, e. g., the required reliability of

8

$1 - 10^{-9}$ for wireless factory automation [Sch+17]. A similar interpretation is proposed in [Pop+18b], defining reliability as the probability that a message arrives before its deadline. In [3GP16b] defines availability as the ratio between the time amount services are delivered and the time amount services are expected with respect to a specific area. The authors of [Xu+08] proposed the terms "service availability" and "communication reliability", highlighting their potentially different timescale. In [SM16], resilience is introduced as ability to ensure the correct functionality for a specified duration, which corresponds to the definition of reliability in traditional dependability theory.

Although 5G and beyond is expected to provide URLLC, which even contains the term reliability, only few research activities have been performed aiming to leverage the respective dependability theory to wireless channels. In [NP16], an analysis framework for evaluating the reliability performance of multi-interface communications is proposed which combines traditional models from dependability theory, such as CTMCs, with technology-specific latency probability distributions. By means of simulations, the authors confirm that series, parallel, and $koon$ models can accurately model the reliability of the considered packet splitting strategies. URLLC is inevitably connected to time-based aspects due to latency deadlines, often in the (sub-)millisecond range, e. g., $250\,\mu s$ for factory automation, $3\,ms$ in smart grids, and $10\,ms$ for intelligent transport systems [Sch+17]. Moreover, zero mobility interruption is targeted [Sim+17], which is obviously the optimal value. [NLP19] analyzes the impact of interface diversity on the burst length of failed and successful packet transmissions, corresponding to the dependability theory metrics downtime and uptime, respectively. The work is based on simulations of a Gilbert-Elliott two-state burst error model, where the transition probability parameters were estimated from latency measurements of Long Term Evolution (LTE) and wireless local area network (WLAN) deployments up to the diversity order of two. The results show that using interface diversity configurations which include at least one WLAN interface leads to superior results in terms of downtime duration.

Besides, time-based dependability metrics, such as MUT, MDT, MTBF, or MTTFF, which are fundamental and well accepted tools in dependability theory, remain almost unmentioned in the research on wireless communications. Thus, further research activities are necessary, interconnecting wireless communications and dependability theory as well as focusing on time aspects in order to refine the discussion on URLLC, which aims for reliably linking systems from different domains, e. g., wireless communications and factory automation.

2.2 Multi-Connectivity in Wireless Communications

Multi-connectivity is used as an umbrella term, referring to approaches, where a user equipment (UE) is simultaneously connected to multiple communication links. These connections can be on the same or on different frequencies, i.e., intra- and inter-frequency multi-connectivity. Among the intra-frequency approaches, coherent and non-coherent demodulation of signals is possible. Transmission points, i.e., base stations (BSs), can be co-located or non-co-located. The following approaches are distinguished.

Single-Frequency Network

Single frequency network (SFN) is an intra-frequency multi-connectivity concept with non-co-located BSs of the same type [Eri01]. The signals, which are sent from multiple BSs are treated similarly to multipath waves, enhancing the signal-to-interference-plus-noise ratio (SINR) at the receiver. Thus, SFNs require coordination among the BSs within the SFN area in creating the signals as well as tight time and frequency synchronization. Literature on SFNs focuses on the selection of the set of BSs which form an SFN, e. g., [Tes+16a], and the improvement of the system performance, e. g., [HZ04]. Moreover, the optimization of the SINR is investigated, e. g., in [Eri01; Kam+09], and mobility robustness is optimized, e. g., in [Tes+16b].

Coordinated Multipoint

Coordinated Multipoint (CoMP) is an intra-frequency multi-connectivity concept, where multiple BSs cooperate in transmitting data to multiple users [Lee+12; Sun+13; ZQL13]. Basic CoMP functionalities have been integrated into LTE targeting the throughput enhancement of cell-edge UEs [3GP11]. Prominent CoMP variants comprise the following [MF11]: In coordinated scheduling/coordinated beamforming, transmit power values and beamforming coefficients are adjusted among multiple BSs based on channel state information (CSI) of multiple users. In joint transmission (JT), multiple BSs send data to a single user who combines the multiple signals coherently. Thus, CoMP relies on tight synchronization, centralized scheduling, and powerful backhaul connections [Hua+10]. Detailed surveys on CoMP are provided in, e. g., [Bar+17; Bas+16; Saw+10]. Many publications focus on CoMP-based performance enhancement and optimization solutions on interference alignment techniques [Nor+15], enhanced inter-cell interference coordination in heterogeneous networks (HetNets) [Mar+16], and joint energy and spectral efficiency optimization [Zha+17]. With respect to 5G, the term multi-transmission and reception points (TRPs) is used: In Release 16, multi-TRPs will be introduced, enabling a kind of CoMP operation [Gho+19].

Carrier Aggregation

The inter-frequency technique Carrier Aggregation (CA) has been introduced in 3GPP
Release 10 and refers to the bundling of multiple smaller frequency bands to a virtual
larger band [Ped+11]. The objective is to increase the system capacity and to facilitate
flexible resource usage. CA is realized at the Medium Access Control (MAC) layer. Thus,
it requires a high-speed, low-latency, fiber-based backhaul and centralized scheduling
[Sim+19]. In order to coordinate inter-cell interference, CA has been applied in LTE
HetNets. A reinforcement learning-based approach was proposed in [SBG13], in which
macro and small cells jointly optimize the system performance.

Dual Connectivity

Dual connectivity (DC) has been defined by 3GPP as a feature in Small Cell Enhancements in LTE Release 12 [3GP15] and refers to an inter-frequency multi-connectivity
technique with two non-co-located BSs, of which one is a small cell. Since the two
carriers are combined at the Packet Data Convergence Protocol (PDCP) layer, the requirements for synchronization and scheduling are more relaxed compared with the
approaches introduced above. The benefits of DC are reduced signaling, mobility robustness, increased UE throughput, and enhanced network energy efficiency through longer
small cell sleep periods as shown, e. g., in [Ros+16; WRP16; PR17].

In the context of 5G, DC has been generalized to multi-radio access technology (RAT)
DC [3GP17a]: UEs can establish connections to two BSs with one BS providing LTE
wireless access and the other one providing NR connectivity. In analogy to LTE DC, the
BSs do not require an ideal backhaul because the data flow split is performed at the
PDCP layer.

5G Multi-Connectivity

Research into 5G multi-connectivity, where a UE is connected to more than two BSs of
the same or different frequencies, has gained momentum. Within the context of 5G,
different architectural solutions and concepts of multi-connectivity have been proposed,
e. g., [MVD16; Rav+16]. The authors of [MVD16] conclude that a data split as in
DC, where data streams are split between multiple carriers before their aggregation
at the user, is not suited to satisfy the reliability requirements of URLLC applications
because the failure detection, rerouting, and retransmission exceed the maximum
latency allowed. There is a strong trend in research to focus on extremely high data
rates by utilizing millimeter wave (mmWave) frequencies provided by multiple BSs,
e. g., [Gio+16], and facilitating highly available transmission by combining multiple

links, e. g., in [Öhm+16a; Öhm+16b]. The communication performance of multi-connectivity is quantified in terms of outage probability and throughput in [Wol+19]. In [Pop+19; NLP18] the term "interface diversity" emphasizes the joint utilization of multiple different communication interfaces, which offers additional degrees of diversity and, thus, can help to fulfill the stringent latency-availability requirements of URLLC. Focusing on a multi-RAT architecture, different packet forwarding schemes are compared regarding their latency performance in [NP16], showing the potential of combining multiple RATs or carrier frequencies in order to enable URLLC.

Most research articles on 5G multi-connectivity consider single-user scenarios. An overview of related work on multi-connectivity for multi-cellular, multi-user systems and basics of the involved matching theory is presented in Section 6.1.

2.3 Summary and Conclusions

Dependability theory is a mathematical framework which provides concepts and methods to analyze and optimize life cycles of technical systems by modeling and reducing failures. Originated from probability theory, the formalization of dependability theory fundamentals started in the 1960s. Basic concepts comprise modeling failures and repairs, redundancy, and accurate terminology. Availability and reliability are main attributes of technical components and systems: Availability denotes the probability of failure-free operation at a time instant, whereas reliability is defined as the probability of failure-free operation throughout a given time interval. Thus, referring to reliability without specifying this interval is no valid statement according to dependability theory. Although the ITU has transferred the definitions of dependability theory to communications, the metrics availability and reliability are often used colloquially and incorrectly in this sector. Consequently, understanding and leveraging the fundamental dependability metrics and their differences contribute to refining the discussion, especially on URLLC, which will help mastering key challenges of future wireless communications systems, e.g., in wireless factory automation and real-time remote control.

Multi-connectivity is expected to be a further key concept in order to enable URLLC in 5G and beyond. This collective term describes system architectures, where multiple communication links to a UE are simultaneously established. Multi-connectivity techniques, such as SFN, CoMP, CA, and DC, were originally developed to increase data rates. However, in the context of 5G, several architectural solutions and concepts of multi-connectivity have been proposed, which target the stringent URLLC requirements, e. g., interface diversity, which denotes the joint utilization of multiple different communication interfaces. The available research on multi-connectivity approaches for URLLC concentrates on addressing the requirements in terms of KPIs, such as outage probability or PLR. Besides latency deadlines, time-related performance metrics, which constitute fundamental building blocks of dependability theory, are only covered to a limited extent in the literature on wireless URLLC. This leaves promising potential for further research.

Based on the discussed state of the art, this book utilizes fundamental dependability theory to foster the understanding and application of relevant terms and concepts for wireless communications. It is demonstrated that a joint consideration of several dependability metrics is of great importance for designing future wireless communications systems. URLLC through multi-connectivity is addressed by providing insights on how to leverage dependability theory in Chapters 3, 4, 5 and how to tackle the challenges of multi-cellular, multi-user systems, i. e., interference and the competition for limited resources, in Chapter 6.

Modeling Causes of Failures 3

The work presented in this chapter was first published in the papers [HSF18b], [Höß+19], and [HSF19a]. The author's personal contributions comprise the analysis of selected causes of failures and the development of the Markov models.

Wireless communications systems can be described with existing terms from dependability theory by modeling channels as repairable components. As introduced in Chapter 2, in dependability theory it is often sufficient to distinguish between only two states: an operating state and a failure state. This applies to each component as well as to the system itself. In the usual notion, "up" is used for a component in an operational state whereas "down" refers to a failure state, i. e., in repair if repairable [HR09]. Interpreting the wireless channel as a repairable item corresponds to the Gilbert-Elliot model, which was created to characterize independent impulsive noise and has been successfully used to analyze error patterns of wireless transmission channels [Gil60; ZRM95]. It can be modeled as a CTMC with a binary state x, illustrated in Fig. 3.1a. The transition rates between the two states are referred to as failure rate λ and repair rate μ. The components can be affected by different causes of failure. This chapter focuses on interference and small-scale fading as two relevant causes of failure for wireless communications [TV05; Öhm17]. Their modeling in terms of dependability theory is developed for a wireless multi-connectivity scenario.

In dependability theory, systems are usually modeled to be composed of several independent components. Introducing redundancy by backup components is a standard approach to improve the dependability of a system. We focus on the case of n parallel components, considering the overall system to be operational if and only if at least one component is operational. This translates to wireless communications as follows: A single user is assumed to be connected to n wireless channels simultaneously. The user's communication is successful if at least 1 out of n wireless channels is operational. This diversity type is known as selection combining [Bre59].

The wireless communications system is modeled as an irreducible, homogeneous CTMC. Subsequently, the focus is set on a single cause of failure, i. e., interference or small-scale fading, before the joint modeling of several causes of failure is developed in Section 3.3. The finite system state j is defined as the number of channels which are currently in the operational state. A system with n channels has $n + 1$ states, including the case that no channel is operational. The system state j is decreased by one whenever an operational channel fails and increased by one when a failed channel becomes operational again,

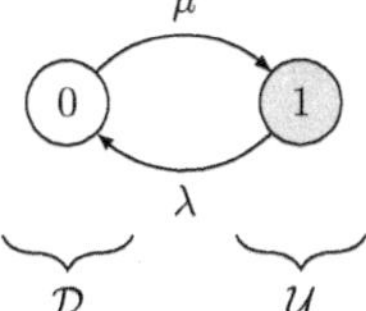

(a) Single channel, correponding to the special case of selection combining with $n = 1$, complies with the Gilbert-Elliot channel model.

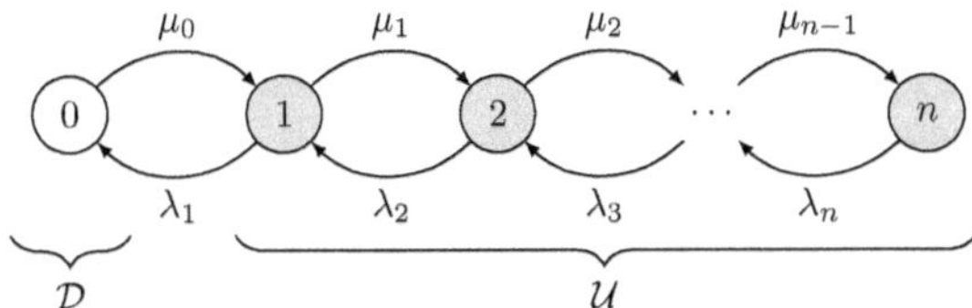

(b) Selection combining of n channels. The system is up, if at least one channel is currently operational.

Fig. 3.1.: CTMC for selection-combined channels with the set of down states $\mathcal{D}$ and up states $\mathcal{U}$ highlighted in green. The state $j = 0, 1, \ldots, n$ reflects the number of channels, which are currently operational.

which corresponds to a channel repair. The state space is divided into the set of "down" states $\mathcal{D} = \{0\}$ and the set of "up" states $\mathcal{U} = \{1, 2, \ldots, n\}$. The resulting CTMC, depicted in Fig. 3.1b, constitutes a birth-death process [Tri16]. The differential balance equations of this CTMC can be expressed in matrix terms as

$$\dot{\boldsymbol{P}}(t) = \boldsymbol{P}(t) \cdot M_{\mathrm{T}}, \tag{3.1}$$

with the state probability row vector $\boldsymbol{P}(t)$, the state probability derivative vector $\dot{\boldsymbol{P}}(t)$, and the tridiagonal $(n+1) \times (n+1)$ transition matrix M_{T}. The system transition parameters comprise the failure rate λ_j and the repair rate μ_j, which characterize transitions originating from each state j. They are derived for the considered wireless communications scenario as

$$\lambda_j = j\lambda \qquad \text{for } 0 \leq j \leq n, \tag{3.2a}$$

$$\mu_j = (n-j)\mu \quad \text{for } 0 \leq j \leq n, \tag{3.2b}$$

based on the assumption of independent channels with equal properties. Thus, failure rates and repair rates in state j are described in terms of the single channel parameters λ and μ, respectively: Since each of the j operational channels may fail, the overall failure rate is increased by the factor j compared to a single channel. In analogy, if one of the $(n-j)$ failed channels recovers, the whole system will be repaired. Hence, the overall repair rate results in an increase of the single channel repair rate μ by the factor $(n-j)$. The corresponding transition matrix is

$$
M_{\mathrm{T}} =
\begin{pmatrix}
-n\mu & n\mu & & & & \\
\lambda & -\lambda-(n-1)\mu & (n-1)\mu & & & \\
& 2\lambda & & \ddots & \ddots & \\
& & \ddots & \ddots & \ddots & \\
& & & \ddots & \ddots & \mu \\
& & & & n\lambda & -n\lambda
\end{pmatrix}
\tag{3.3}
$$

omitting elements equal to 0 for readability.

The following sections provide details on how to model different causes of failures for wireless communications systems, i. e., interference or/and small-scale fading, with respect to the introduced framework.

3.1 Interference

Co-channel interference is an important wireless access issue, especially for systems which operate in unlicensed frequency bands. We assume a single interferer which jams the signal completely, i. e., a user is not able to successfully transmit and receive messages via a channel once the interferer begins to transmit over the same channel. Thus, we introduce the binary random variable *interference state* of channel c_i according to

$$
x_i^{\mathrm{I}}(t) =
\begin{cases}
0, & \text{if channel } c_i \text{ is interfered at } t, \text{ ``down''} \\
1, & \text{if channel } c_i \text{ is not interfered at } t, \text{ ``up''}
\end{cases}
\tag{3.4}
$$

with $i = 1, 2, \ldots, n$. The interferer's arrival rate on a channel c_i is denoted by $\lambda_{i,\mathrm{I}}$. The interferer's service rate is described by $\mu_{i,\mathrm{I}}$. From the user's viewpoint the channel failure and repair rate with respect to interference correspond to $\lambda_{i,\mathrm{I}}$ and $\mu_{i,\mathrm{I}}$, respectively. We assume equal and known probabilities for interferers to appear or leave any channel at all times, which leads to constant interference failure and repair rates, i. e., $\lambda_{i,\mathrm{I}}(t) = \lambda_{\mathrm{I}}$ and $\mu_{i,\mathrm{I}}(t) = \mu_{\mathrm{I}}$, respectively. This knowledge could be estimated and continuously updated based on spectrum sensing. Hence, every interference event is assumed to be self-revealing, i. e., every state change is assumed to be recognized immediately. The probability that more than one channel is blocked or released at the same time is negligible and, thus, no state can be skipped.

3.2 Small-Scale Fading

Small-scale fading characterizes the rapid fluctuations of the amplitudes, phases, or multipath delays of a radio signal over a short period of time or travel distance [Rap02]. A transmitted signal is reflected from surrounding objects, and multiple versions, each with different amplitude and phase, arrive at the receiver. Their superposition potentially leads to constructive or destructive interference, which results in amplified or attenuated signal power. In case of strong attenuation, the signal experiences so-called deep fades. Small-scale fading poses a major challenge for wireless communications systems because the received signal can degrade by several orders of magnitude due to small changes in position, which potentially disrupts connectivity. The variation of small-scale fading in time is caused by the Doppler effect: due to the relative motion between the mobile and the base station, each multipath wave experiences an apparent shift in frequency [Rap02].

In this book, a set of n channels is considered, which are separated in frequency at least by the coherence bandwidth, resulting in independent small-scale fading. Compensation of path loss and shadowing by transmit power or automatic gain control is assumed [ÖF15]. However, the contributions of this work can be extended to wireless systems that include path loss and shadowing as well as other channel assumptions by determining the respective failure and repair rates. It is assumed that a channel is operational if and only if messages can be successfully transmitted and received over it. As a simplification, it is assumed that a signal subjected to fading can be successfully received if the instantaneous power $p_i(t)$ of a channel $i \in \{1, 2, \ldots, n\}$ is above a certain threshold $p_{\min}$, which may be determined by the hardware sensitivity of the receiver. Thus, we distinguish between two states by introducing the random variable *fading state* of channel c_i according to

$$x_i^{\mathrm{F}}(t) = \begin{cases} 0, & \text{if } p_i(t) < p_{\min}, \text{ channel } c_i \text{ faded, "down"} \\ 1, & \text{if } p_i(t) \geq p_{\min}, \text{ channel } c_i \text{ not faded, "up"} \end{cases} \tag{3.5}$$

applying the same threshold $p_{\min}$ for each channel. The n considered channels are assumed to experience the same average receive power p_{avg}. The fading margin is specified by

$$\Phi = \frac{p_{\mathrm{avg}}}{p_{\min}}, \tag{3.6}$$

which can be interpreted as normalized signal-to-noise ratio (SNR) because the power threshold $p_{\min}$ can be associated with a minimum SNR for a given noise power. In the following, two prominent types of small-scale fading are considered, namely Rayleigh and Rice fading.

3.2.1 Rayleigh Fading

Dynamic multi-path propagation assuming each wireless channel to consist of multiple paths without a dominant component leads to the Rayleigh fading channel [TV05]. It is an idealized model for a time-variant transmission channel, which is often for mobile communication links under non-line-of-sight (NLOS) conditions. The following assumptions apply within the Rayleigh fading channel model [Nus14].

Level-crossing analysis provides the average non-outage and outage duration of a Rayleigh-faded signal. Their reciprocals characterize the fading failure rate $\lambda_\mathrm{F}^\mathrm{Ray}$ and fading repair rate $\mu_\mathrm{F}^\mathrm{Ray}$, respectively, according to [ÖF15]

$$\lambda_\mathrm{F}^\mathrm{Ray} = \sqrt{\frac{2\pi}{\Phi}}\, f_\mathrm{D}, \tag{3.7a}$$

$$\mu_\mathrm{F}^\mathrm{Ray} = \frac{\sqrt{\frac{2\pi}{\Phi}}\, f_\mathrm{D}}{\exp\frac{1}{\Phi} - 1}. \tag{3.7b}$$

The maximum Doppler frequency is represented by $f_\mathrm{D} = vf/c$ with f denoting the carrier frequency of the signal and c the speed of light. The relative velocity between transmitter, receiver, and scatterers is characterized by v. The random fading process is assumed to be stationary. Thus, the transition rates between the two fading states, denoted as fading failure rate $\lambda_\mathrm{F}^\mathrm{Ray}$ and fading repair rate $\mu_\mathrm{F}^\mathrm{Ray}$, are constant and equal for any channel. Further assumptions are as follows: Every fading is self-revealing, i.e., every state change is recognized immediately. The probability that more than one channel enters or leaves the failure state at exactly the same time is negligible.

3.2.2 Rice Fading

A channel consisting of many individual paths including a dominant component, e. g., in line-of-sight (LOS) scenarios, corresponds to a Rice fading channel [Abd+00]. The Rice K-factor specifies the power ratio between the dominant and non-dominant components. Thus, the Rayleigh fading channel is a special case of the Rice fading channel with $K = 0$. The further assumptions of the previous section still apply.

In analogy to the procedure for the Rayleigh fading channel, level-crossing analysis determines the average non-outage and outage duration of a Rice-faded signal, whose reciprocals specify the failure and repair rate, respectively. The level-crossing rate of a Rice-faded signal is given by [Abd+00]

$$N_\mathrm{R} = \frac{1}{\bar{\tau}^\mathrm{u} + \bar{\tau}^\mathrm{d}} \tag{3.8}$$

$$= \sqrt{\frac{2\pi(K+1)}{\Phi}}\, f_\mathrm{D} \exp\left(-K - \frac{K+1}{\Phi}\right) \cdot I_0\left(2\sqrt{\frac{K(K+1)}{\Phi}}\right), \tag{3.9}$$

where $I_0(\cdot)$ denotes the zero-order modified Bessel function of the first kind. By incorporating the known average fade duration $\bar{\tau}^\mathrm{d}$ [Abd+00], we determine the fading failure and repair rate according to

$$\lambda_\mathrm{F} = \frac{1}{\bar{\tau}^\mathrm{u}} = \frac{N_\mathrm{R}}{Q\left(\sqrt{2K}, \sqrt{\frac{2(K+1)}{\Phi}}\right)}, \tag{3.10a}$$

$$\mu_\mathrm{F} = \frac{1}{\bar{\tau}^\mathrm{d}} = \frac{N_\mathrm{R}}{1 - Q\left(\sqrt{2K}, \sqrt{\frac{2(K+1)}{\Phi}}\right)} \tag{3.10b}$$

with $Q(\cdot, \cdot)$ denoting the Marcum Q-function. Applying $K = 0$ yields the known level-crossing rate expression of a Rayleigh-faded signal [Rap02]

$$N_\mathrm{R}^\mathrm{Ray} = \lim_{K \to 0} N_\mathrm{R} = \sqrt{\frac{2\pi}{\Phi}}\, f_\mathrm{D} \exp\left(-\frac{1}{\Phi}\right) \tag{3.11}$$

and the corresponding fading failure and repair rates for Rayleigh fading (3.7) because $Q(0, y) = \exp\left(\frac{-y^2}{2}\right)$ [PS08].

3.3 Interference & Small-Scale Fading

Subsequently, we jointly consider both causes of failure, interference and small-scale fading, for a single channel and two selection-combined channels. Taking into account several causes of failure simultaneously indicates that the channels, which constitute the system, are modeled to be composed of repairable components themselves. This approach requires the extension of the introduced CTMC.

Tab. 3.1.: State definitions

J	0	1	2	3	4	5	6	7	8	9	10	11	12	13	14	15
x_1^{I}	0	1	0	1	0	1	0	1	0	1	0	1	0	1	0	1
x_1^{F}	0	0	1	1	0	0	1	1	0	0	1	1	0	0	1	1
x_2^{I}	0	0	0	0	1	1	1	1	0	0	0	0	1	1	1	1
x_2^{F}	0	0	0	0	0	0	0	0	1	1	1	1	1	1	1	1

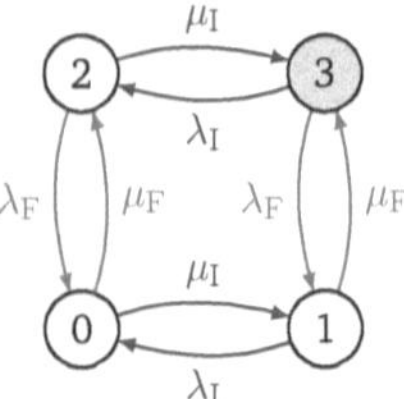

Fig. 3.2.: CTMC for one channel with interference and small-scale fading.

3.3.1 Single Channel with Interference and Small-Scale Fading

A single channel c_i is modeled to comprise two components: an interference stage x_i^{I} and a fading stage x_i^{F}. The overall state of a single channel c_i can be described by a binary function

$$x_i = \phi\left(x_i^{\mathrm{I}}, x_i^{\mathrm{F}}\right) = \begin{cases} 0, & \text{if channel } c_i \text{ is failed, "down"} \\ 1, & \text{if channel } c_i \text{ is operational, "up"} \end{cases}. \tag{3.12}$$

We refer to $\phi\left(x_i^{\mathrm{I}}, x_i^{\mathrm{F}}\right)$ as the structure function of a single channel. If and only if messages can be successfully transmitted and received via a channel, the latter is assumed to be operational. Thus, it is obvious that the channel is up if and only if all components are up. This corresponds to a *series structure* with [HR09]

$$x_i = \phi\left(x_i^{\mathrm{I}}, x_i^{\mathrm{F}}\right) = x_i^{\mathrm{I}} \cdot x_i^{\mathrm{F}}. \tag{3.13}$$

We model the considered wireless communications system as an irreducible, homogeneous CTMC. In this regard, we fix $i = 1$ without loss of generality. Let the finite system state J be defined as the decimal representation of $\left(x_1^{\mathrm{F}}, x_1^{\mathrm{I}}\right)$. Hence, a system with one channel has $n_1 = 4$ states, which can be found in the first four columns of Table 3.1. The state space is partitioned into the set of "up" states $\mathcal{U}_1 = \{3\}$ (highlighted green) and the set of "down" states $\mathcal{D}_1 = \{0, 1, 2\}$. The resulting CTMC is depicted in Fig. 3.2. The differential balance equations of this CTMC can be expressed in matrix terms as

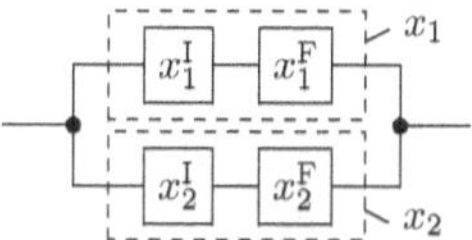

Fig. 3.3.: Reliability block diagram of two parallel channels with interference and small-scale fading.

$\dot{\boldsymbol{P}}(t) = \boldsymbol{P}(t) \cdot M_{\mathrm{T1}}$ with the state probability row vector $\boldsymbol{P}(t)$, the state probability derivative vector $\dot{\boldsymbol{P}}(t)$, and the 4×4 transition matrix

$$M_{\mathrm{T1}} = \begin{pmatrix} -\mu_{\mathrm{I}} - \mu_{\mathrm{F}} & \mu_{\mathrm{I}} & \mu_{\mathrm{F}} & 0 \\ \lambda_{\mathrm{I}} & -\lambda_{\mathrm{I}} - \mu_{\mathrm{F}} & 0 & \mu_{\mathrm{F}} \\ \lambda_{\mathrm{F}} & 0 & -\lambda_{\mathrm{F}} - \mu_{\mathrm{I}} & \mu_{\mathrm{I}} \\ 0 & \lambda_{\mathrm{F}} & \lambda_{\mathrm{I}} & -\lambda_{\mathrm{F}} - \lambda_{\mathrm{I}} \end{pmatrix}. \tag{3.14}$$

3.2 Two Channels with Interference and Small-Scale Fading

Introducing redundancy by an additional component corresponds to selection combining of two channels. The system will be operational if and only if at least one of the components is operational, i.e., the user's communication is successful if at least one out of two wireless channels c_1, c_2 is operational. The state of the whole system can be characterized by the binary structure function

$$\theta\left(x_1, x_2\right) = \begin{cases} 0, & \text{if the system is failed, "down"} \\ 1, & \text{if the system is operational, "up"} \end{cases} \tag{3.15}$$

with x_1, x_2 denoting the states of the two individual channels c_1, c_2, respectively. This translates to a *parallel structure* with [HR09]

$$\theta\left(x_1, x_2\right) = 1 - (1 - x_1)(1 - x_2). \tag{3.16}$$

Inserting the structure function (3.13) of both channels c_1, c_2 with their interference and fading components $x_1^{\mathrm{I}}, x_1^{\mathrm{F}}, x_2^{\mathrm{I}}, x_2^{\mathrm{F}}$ yields

$$\theta\left(\phi\left(x_1^{\mathrm{I}}, x_1^{\mathrm{F}}\right), \phi\left(x_2^{\mathrm{I}}, x_2^{\mathrm{F}}\right)\right) = 1 - \left(1 - x_1^{\mathrm{I}} x_1^{\mathrm{F}}\right)\left(1 - x_2^{\mathrm{I}} x_2^{\mathrm{F}}\right). \tag{3.17}$$

This can be illustrated by the reliability block diagram[1] in Fig. 3.3 showing the logical connections of components necessary to fulfill the system function.

---

[1] The term "reliability block diagram" is the usual term for a success-oriented network describing a binary function of the system [HR09]. It is not limited to the KPI reliability (cf. Chapter 2) but originates from the analysis of non-repairable systems, where the dependability metrics reliability and availability coincide.

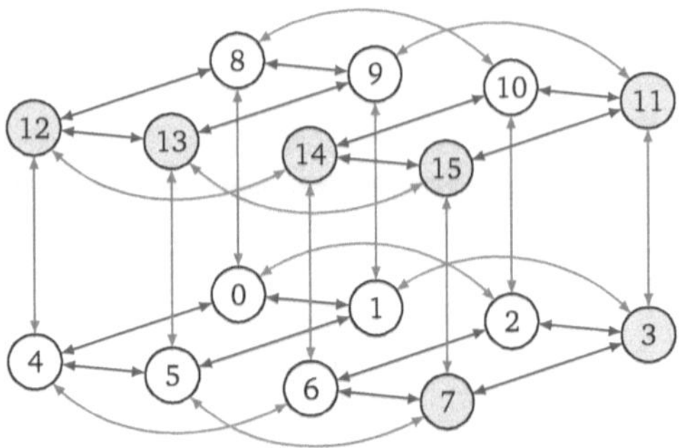

Fig. 3.4.: CTMC for two selection-combined channels with interference and small-scale fading.

The CTMC introduced in the previous section is extended by adding the second channel. The finite system state J is defined as the decimal representation of $\left(x_2^{\mathrm{F}}, x_2^{\mathrm{I}}, x_1^{\mathrm{F}}, x_1^{\mathrm{I}}\right)$ according to Table 3.1. The system with two channels comprises $n_2 = 16$ states. The set of "up" states results in $\mathcal{U}_2 = \{3, 7, 11, 12, 13, 14, 15\}$ (highlighted in green) and the set of "down" states is $\mathcal{D}_2 = \{0, 1, 2, 4, 5, 6, 8, 9, 10\}$. The obtained CTMC is shown in Fig. 3.4. For readability, the state transition rates are omitted. In analogy to Fig. 3.2, blue (red) transitions characterize interference (fading) state changes. Transitions to a higher (lower) system state J refer to a repair (failure). The 16×16 transition matrix results in

$$
M_{\mathrm{T2}} = \begin{pmatrix}
\cdots & \mu_{\mathrm{I}} & \mu_{\mathrm{F}} & & \mu_{\mathrm{I}} & & & & \mu_{\mathrm{F}} & & & & & & & \\
\lambda_{\mathrm{I}} & \cdots & & \mu_{\mathrm{F}} & & \mu_{\mathrm{I}} & & & & \mu_{\mathrm{F}} & & & & & & \\
\lambda_{\mathrm{F}} & & \cdots & \mu_{\mathrm{I}} & & & \mu_{\mathrm{I}} & & & & \mu_{\mathrm{F}} & & & & & \\
& \lambda_{\mathrm{F}} & \lambda_{\mathrm{I}} & \cdots & & & & \mu_{\mathrm{I}} & & & & \mu_{\mathrm{F}} & & & & \\
\lambda_{\mathrm{I}} & & & & \cdots & \mu_{\mathrm{I}} & \mu_{\mathrm{F}} & & & & & & \mu_{\mathrm{F}} & & & \\
& \lambda_{\mathrm{I}} & & & \lambda_{\mathrm{I}} & \cdots & & \mu_{\mathrm{F}} & & & & & & \mu_{\mathrm{F}} & & \\
& & \lambda_{\mathrm{I}} & & \lambda_{\mathrm{F}} & & \cdots & \mu_{\mathrm{I}} & & & & & & & \mu_{\mathrm{F}} & \\
& & & \lambda_{\mathrm{I}} & & \lambda_{\mathrm{F}} & \lambda_{\mathrm{I}} & \cdots & & & & & & & & \mu_{\mathrm{F}} \\
\lambda_{\mathrm{F}} & & & & & & & & \cdots & \mu_{\mathrm{I}} & \mu_{\mathrm{F}} & & \mu_{\mathrm{I}} & & & \\
& \lambda_{\mathrm{F}} & & & & & & & \lambda_{\mathrm{I}} & \cdots & & \mu_{\mathrm{F}} & & \mu_{\mathrm{I}} & & \\
& & \lambda_{\mathrm{F}} & & & & & & \lambda_{\mathrm{F}} & & \cdots & \mu_{\mathrm{I}} & & & \mu_{\mathrm{I}} & \\
& & & \lambda_{\mathrm{F}} & & & & & & \lambda_{\mathrm{F}} & \lambda_{\mathrm{I}} & \cdots & & & & \mu_{\mathrm{I}} \\
& & & & \lambda_{\mathrm{F}} & & & & \lambda_{\mathrm{I}} & & & & \cdots & \mu_{\mathrm{I}} & \mu_{\mathrm{F}} & \\
& & & & & \lambda_{\mathrm{F}} & & & & \lambda_{\mathrm{I}} & & & \lambda_{\mathrm{I}} & \cdots & & \mu_{\mathrm{F}} \\
& & & & & & \lambda_{\mathrm{F}} & & & & \lambda_{\mathrm{I}} & & \lambda_{\mathrm{F}} & & \cdots & \mu_{\mathrm{I}} \\
& & & & & & & \lambda_{\mathrm{F}} & & & & \lambda_{\mathrm{I}} & & \lambda_{\mathrm{F}} & \lambda_{\mathrm{I}} & \cdots
\end{pmatrix}
\tag{3.18}
$$

with zeros omitted. The diagonal elements, which are abbreviated as dots, equal the negative sum over the corresponding row entries, which result from the balance equations for Markov processes [HR09]. Due to the fact that more than one cause of failure is

modeled, the state space of the CTMC grows exponentially with the number of channels
and the number of causes of failures because all combinations of up and down states are
covered. An approach to reduce complexity is state aggregation [Kle+16], e. g., into one
up state and one down state because this information is of actual interest. However, state
aggregation suffers from inaccuracies because information on the dynamics within the
aggregated states are lost (cf. [Sch20]). Thus, state aggregation will not be employed in
the context of this book.

3.4 Summary and Conclusions

By applying dependability theory to wireless communications, this chapter builds the
foundation for the subsequent evaluations. It is demonstrated how to leverage the
existing toolset of dependability theory, e. g., Markov analysis, structure functions,
and reliability block diagrams, to the communications domain. A simplified wireless
communications system utilizing selection combining is modeled as a system comprising
repairable components, represented by a CTMC. Two potential causes of failure are
considered individually as well as jointly, i. e., co-channel interference and small-scale
fading based on Rayleigh and Rice fading. The state space is defined and the central
parameters failure rate and repair rate are derived. For a single cause of failure, a
birth-death CTMC is obtained, whose number of states is proportional to the number
of channels. If two causes of failure are studied jointly, the number of states grows
exponentially with the diversity order. This is the reason why this work concentrates on
selection combining with two channels, i. e., dual connectivity. The introduced CTMC
allows the dependability evaluation and improvement of multi-connectivity regarding
average and time-related performance, as well as success probabilities throughout
mission intervals, developed in the subsequent chapter.

Improving Wireless Dependability through Selection Combining

4

The work presented in this chapter was first published in the papers [Höß+17], [HSF18a], [HSF18b], [HSF19a], [Höß+19], [Tra+19a] and [Tra+19b]. The author's personal contributions comprise the selection and derivation of KPIs, implementation and execution of the simulations, and evaluation of the resulting data.

The realization of wireless URLLC is one of the key challenges of 5G and beyond. Ensuring latency in the (sub-)millisecond range together with ultra-high reliability is expected to enable wireless factory automation, self-driving cars, and real-time remote control, paving the way to the Tactile Internet [Sim+16; Sch+17]. The wireless communications community usually considers reliability as percentage of successful packet delivery, aiming for $1-10^{-5}$ up to $1-10^{-9}$ in URLLC [Sch+17; BDP18; 3GP17c]. However, by applying fundamental dependability theory this chapter will show that the commonly used metric meets the definition of availability instead of reliability and it fails in reflecting any dependency on the time dimension. Although any wireless communications system suffers from signal fluctuations over time, referred to as fading, analysis of time-based dependability attributes is lacking in the context of URLLC. According to [Sim+17], zero mobility interruption is targeted for URLLC categories, which is the only explicit time attribute in the current discussion. However, this optimal value cannot be guaranteed with $100\,\%$ certainty due to the random fading in wireless channels.

This chapter utilizes dependability theory to contribute to the realization of wireless URLLC by examining appropriate KPIs. Since diversity is widely considered to be key in order to compensate for fading, e. g., by sending messages redundantly over different wireless channels [BDP18], potential dependability gains due to multi-connectivity are discussed for selection-combined channels. The focus is set on selection combining is because of its low complexity compared to other multi-connectivity schemes, e. g., maximum-ratio combining or joint decoding [Wol+19]. We demonstrate the benefit of time-related and in particular mission-related KPIs, which, as elaborated in Chapter 2, remain almost unmentioned in wireless communications although they have been well accepted for many years in other areas, such as electronic systems design [BP65; HR09]. Basic dependability quantities will be introduced and applied to the considered wireless

24

communications system, which is modeled as a repairable system corresponding to Chapter 3. Closed-form expressions for mean up- and downtimes of the diversity system under Rayleigh fading will be determined and extended to the more general Rice fading, which allows to cover also channels including a dominant path, i.e., a LOS component. Moreover, we jointly study two causes of failure, i.e., interference and small-scale fading. Alongside the discussion of dependability metrics which focus on the average performance, the probability distributions of up- and downtimes are evaluated, in order to enable concrete performance guarantees. By introducing the metric "mission reliability", we will present how the success probability of a continuous mission, which does not allow a failure, depends on its mission duration. In addition, we study the success probability of a mission if the system is able to tolerate short interruptions, assuming that today's communication systems can tolerate short outage times.

4.1 Average and Time-Related Dependability KPIs

Important quantities used in dependability theory focus on average success probabilities and time duration. In this section several of those dependability metrics are derived for the considered wireless communications system. According to the system model introduced in Chapter 3, we identify the repairable components of a system with wireless channels. The general concept of $koon$ redundancy can be applied, which characterizes a system that is functioning if and only if at least k of the n channels are operational [HR09]. In the introduced CTMC which considers a single cause of failure, the set of down states is $\mathcal{D} = 0, 1, \ldots, k - 1$ and the set of up states results in $\mathcal{U} = k, k + 1, \ldots, n$. Hence, the parameter k corresponds to the number of down states $k = |\mathcal{D}|$, as illustrated in Fig. 3.1b for $k = 1$. In the following, closed form expressions of average and time-related dependability KPIs are presented for this special case $1oon$ redundancy, equivalent to selection combining of n channels. In this context, the index n relates to the number of channels a user is simultaneously connected to, performing selection combining.

4.1.1 Channel Availability

According to the ITU [ITU08], "an item is available, if it is in a state to perform a required function at a given instant of time or at any instant of time within a given time interval, assuming that the external resources, if required, are provided." On the basis of this definition the following availability quantities can be derived with respect to the wireless channel.

Definition 1. *The **instantaneous channel availability***

$$A(t) = \Pr\left\{x(t) = 1\right\} \tag{4.1}$$

is the probability that a channel is operational at a given instant of time t.

Since it focuses on a point in time, it is also called *point availability* [Bir13].

Definition 2. *The **steady-state channel availability***

$$A = \lim_{t \to \infty} A(t) \tag{4.2}$$

characterizes the long-term probability that one channel is operational.

The steady-state channel availability can also be interpreted as the mean proportion of time the channel is operational. In the following, steady-state channel availability is also simply referred to as channel availability.

We can apply the concept of availability to the introduced scenario, because the considered wireless communications system is available if it is in one of the system up states aggregated in $\mathcal{U}$. The steady-state situation is often of special interest to make conclusions about the system's average performance. Thus, we determine the system steady-state channel availability for selection combining of n channels as

$$A_n = \sum_{j \in \mathcal{U}} P_j = \sum_{j=k}^{n} P_j \, . \tag{4.3}$$

The steady-state probabilities $\boldsymbol{P} = [P_0, P_1, \dots, P_n]$ satisfy the matrix equation

$$\boldsymbol{P} \cdot M_\mathrm{T} = [0, 0, \dots, 0] \tag{4.4}$$

due to the steady-state condition

$$\dot{\boldsymbol{P}} = 0. \tag{4.5}$$

The complementary steady-state channel availability characterizes the outage probability given by

$$P_n^\mathrm{out} = 1 - A_n = \sum_{j \in \mathcal{D}} P_j = \sum_{j=0}^{k-1} P_j \, . \tag{4.6}$$

It corresponds to the PLR, a quantity often used to specify dependability requirements in communications systems, because it can be interpreted as the long-term probability that the communications system is not operational.

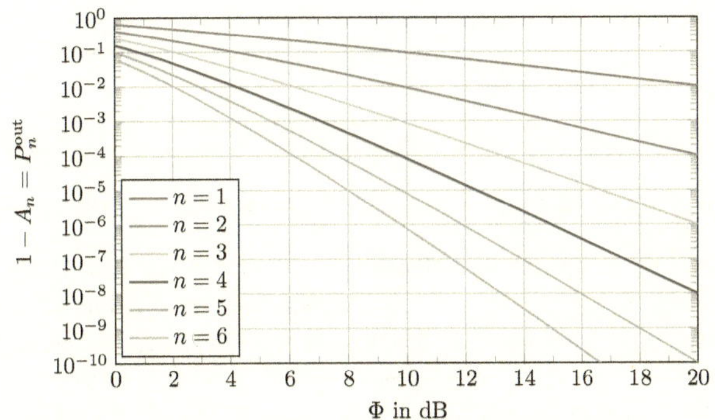

Fig. 4.1.: Complementary steady-state channel availability for n selection-combined Rayleigh fading channels depending on the fading margin Φ.

Channel Availability of Selection-Combined Rayleigh Fading Channels

Focusing on Rayleigh fading as a single cause of failure, the considered system model is a birth-death Markov process, whose steady-state probabilities can be computed by [Tri16]

$$P_j = \frac{\prod_{i=0}^{j-1}\left(\frac{\mu_i}{\lambda_{i+1}}\right)}{1 + \sum_{\ell=1}^{n}\prod_{m=0}^{\ell-1}\left(\frac{\mu_m}{\lambda_{m+1}}\right)} \tag{4.7}$$

for $j = 0, 1, \ldots, n$. We apply the transition rates (3.2) according to the introduced wireless communications system scenario, obtaining

$$P_j = \frac{n!}{j!(n-j)!\rho^j}\left(1 + \sum_{\ell=1}^{n}\frac{n!}{\ell!(n-l)!\rho^\ell}\right)^{-1} \tag{4.8}$$

with the ratio

$$\rho = \lambda/\mu = \exp\left(\frac{1}{\Phi}\right) - 1 . \tag{4.9}$$

For readability, the subscript "F" and superscript "Ray" at transition rates are waived if Rayleigh fading is considered. It is obvious that, for the considered wireless communications system scenario, the steady-state probabilities and, consequently, the user's steady-state channel availability purely depend on the fading margin Φ. Hence, this metric does not reflect the influence of mobility aspects or the carrier frequency on the communication performance.

Applying (4.3) for the special case $k = 1$, which holds for the entire chapter, leads to the steady-state channel availability of n selection-combined Rayleigh fading links

$$A_n = 1 - \frac{\lambda^n}{(\lambda + \mu)^n} = 1 - \left(1 - \exp\left(-\frac{1}{\Phi}\right)\right)^n \tag{4.10}$$

confirming the known probability expression for Rayleigh fading channels [ÖF15].

Evaluations of the complementary steady-state channel availability A_n are shown for different values of Φ and n in Fig. 4.1. Higher degrees of redundancy, which are equivalent to higher numbers n of selection-combined channels, improve the steady-state channel availability. Increasing fading margins Φ cause higher gains in terms of steady-state channel availability for systems with a larger number n of channels. Interpreting outage probability P_n^{out} as PLR, as assumed in this work, enables system design recommendations for a given fading margin Φ. For example in this scenario, it is not possible to achieve a PLR $< 10^{-8}$ for a user connected to 4 links simultaneously in the plotted range of Φ.

Channel Availability Regarding Rice Fading and Interference

If interference (I) and Rice fading (F) are considered as potential causes of failure according to the system model presented in Section 3.3, the following availability quantities can be derived for the individual components of a wireless channel c_i:

$$A^{\text{I}} = A_i^{\text{I}} = \Pr\left\{x_i^{\text{I}}\right\}, \tag{4.11}$$

$$A^{\text{F}} = A_i^{\text{F}} = \Pr\left\{x_i^{\text{F}}\right\}, \tag{4.12}$$

$$A_1 = \Pr\left\{x_i\right\} = \Pr\left\{x_i^{\text{I}} \wedge x_i^{\text{F}}\right\} = A^{\text{I}} A^{\text{F}}, \tag{4.13}$$

$$A_2 = \Pr\left\{x_1 \vee x_2\right\} = 1 - \Pr\left\{\bar{x}_1 \wedge \bar{x}_2\right\} = 1 - (1 - A_1)^2 = 1 - (1 - A^{\text{I}} A^{\text{F}})^2. \tag{4.14}$$

with $i \in \{1, 2\}$ and assuming the same statistics for both channels. A_1 denotes the availability of one channel, whereby A_2 characterizes the availability of the whole system comprising two selection-combined channels. The symbols $\wedge$ and $\vee$ describe the logical operations conjunction and disjunction, respectively. Since availability corresponds to the mean proportion of time a component (or system) is operational, we derive the availability of the fading component according to

$$A^{\text{F}} = \frac{\bar{\tau}^{\text{u}}}{\bar{\tau}^{\text{u}} + \bar{\tau}^{\text{d}}} = \frac{\mu_{\text{F}}}{\mu_{\text{F}} + \lambda_{\text{F}}} = Q\left(\sqrt{2K}, \sqrt{\frac{2(K + 1)}{\Phi}}\right) = 1 - P^{\text{F,out}}, \tag{4.15}$$

which complies with the known expression for a Rice fading channel [Abd+00]. It is worth mentioning that this fading availability purely depends on the fading margin Φ and the Rice K-factor. As in the previous section on Rayleigh fading, the availability does not

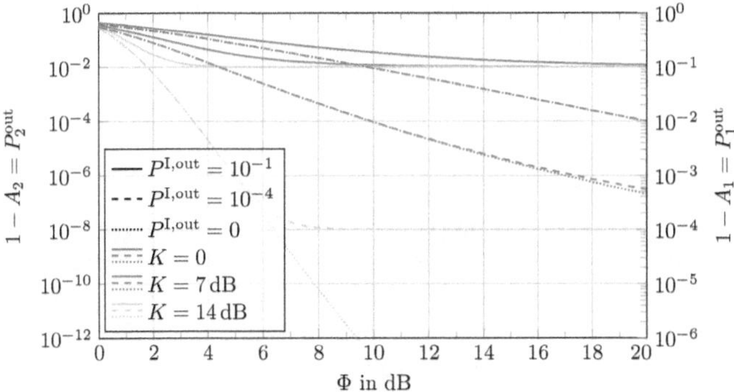

Fig. 4.2.: Complementary steady-state channel availability for $n \in \{1,2\}$ channel(s) depending on the fading margin Φ for different Rice K-factors and different interference outage probabilities $P^{\mathrm{I,out}}$.

reflect the influence of mobility aspects or the carrier frequency on the communication performance. The availability of the component modelling interference yields

$$A^{\mathrm{I}} = \frac{\mu_{\mathrm{I}}}{\mu_{\mathrm{I}} + \lambda_{\mathrm{I}}} = \frac{1}{1 + \frac{\lambda_{\mathrm{I}}}{\mu_{\mathrm{I}}}} = 1 - P^{\mathrm{I,out}}. \tag{4.16}$$

Obviously, all interference failure and repair rates with the same ratio lead to the same interference availability value A^{I}.

Alternatively, the channel availabilities for one and two channels, A_1, A_2, can be determined by utilizing the CTMCs introduced in Section 3.3. The considered wireless communications system is available if it is in one of the system up states aggregated in $\mathcal{U}_1$ and $\mathcal{U}_2$, respectively. Solving (4.3) and (4.4) results in [HSF19a]

$$A_n = 1 - \left(1 - A^{\mathrm{I}} A^{\mathrm{F}}\right)^n = 1 - P_n^{\mathrm{out}}, \tag{4.17}$$

which is consistent with the obtained expressions (4.13) and (4.14) for $n \in \{1,2\}$.

Fig. 4.2 depicts the complementary steady-state channel availability $1 - A_n$, equivalent to the outage probability, for $n \in \{1,2\}$ channel(s). Consistent with the previous discussion on selection combining with pure Rayleigh fading, adding a parallel channel leads to exponentially improved availability, $P_2^{\mathrm{out}} = \left(P_1^{\mathrm{out}}\right)^2$, which follows from (4.13) and (4.14). Thus, both cases are presented in one plot, utilizing two different y-axes. The outage probabilities decrease with higher fading margins Φ. The overall availability within the series structure of a single channel is limited by the component (interference or fading) which exhibits the higher outage probability. Higher fading margins lead to improved availability of the fading component with $\lim_{\Phi \to \infty} A^{\mathrm{F}} = 1$, implying that the overall availability approaches the availability of the interference component: Determining the limit of (4.17) results in

$$\lim_{\Phi \to \infty} A_n = 1 - \left(1 - A^{\mathrm{I}}\right)^n = 1 - \left(P^{\mathrm{I,out}}\right)^n. \tag{4.18}$$

This effect is clearly visible in Fig. 4.2 because P_n^{out} approaches $\left(P^{\mathrm{I,out}}\right)^n$ with $n \in \{1, 2\}$ for increasing fading margin Φ. Higher Rice K-factors, which indicate more dominant LOS components in the signal, can reduce the system outage probability by several orders of magnitude. The gain increases for larger fading margins Φ. Equivalently, this translates to a gain in terms of fading margin for a given outage probability: Comparing $K = 7\,\mathrm{dB}$ and $K = 14\,\mathrm{dB}$ corresponds to a fading margin gain of $12\,\mathrm{dB}$ for $P_2^{\mathrm{out}} = 10^{-6}$, which is considered sufficient for many URLLC applications [Sch+19b].

However, it is obvious that this KPI fails to reflect the performance with respect to time-related aspects, e. g., time-varying channels or the duration of an interference state in a wireless system: For instance, an availability value of $99\,\%$ may refer to frequent failures (e. g., on average every $99\,\mathrm{ms}$) and short downtimes ($1\,\mathrm{ms}$) as well as long downtimes (e. g., $10\,\mathrm{min}$) and relatively rare failures (one failure every $16.5\,\mathrm{h}$). The consequences for a real system are of course fundamentally different. Thus, it is beneficial to define requirements characterizing the duration of uptime, downtime, and time between failures.

4.1.2 Mean Time Between Failures

The MTBF is a central KPI characterizing a time-related dependability aspect, defined as follows [HR09].

Definition 3. *The **mean time between failures** MTBF is the average time duration between consecutive transitions from an up state to a down state.*

Due to the renewal process of a repairable system, the MTBF_n is constituted by the sum of the mean uptime MUT_n and mean downtime MDT_n,

$$\mathrm{MTBF}_n = \mathrm{MUT}_n + \mathrm{MDT}_n. \tag{4.19}$$

The KPIs MDT_n and MUT_n will be discussed in the sections 4.1.3 and 4.1.4, respectively.

Based on the frequency of system failures

$$\omega_n = \sum_{i \in \mathcal{U}} \sum_{\ell \in \mathcal{D}} P_i \cdot a_{i\ell}, \tag{4.20}$$

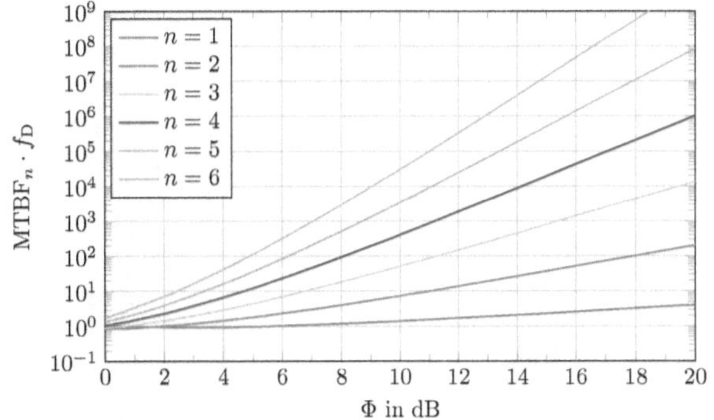

Fig. 4.3.: MTBF for selection-combined Rayleigh fading channels and varying values of the fading margin Φ and the number of selection-combined channels n.

the MTBF$_n$ is generally determined as

$$\mathrm{MTBF}_n = \frac{1}{\omega_n},\qquad(4.21)$$

where $a_{i\ell}$ denotes the transition rate from state i to ℓ [HR09]. Subsequently, this definition is evaluated for the introduced wireless communications scenario considering Rayleigh fading.

For n selection-combined Rayleigh fading channels, the MTBF expression simplifies to

$$\mathrm{MTBF}_n = \frac{1}{\lambda P_1} = \frac{(\lambda + \mu)^n}{n\mu\lambda^n}\qquad(4.22)$$

due to the birth-death structure of the considered system model, applying the steady-state probability (4.7) with $k = 1$. This can be rewritten as

$$\mathrm{MTBF}_n = \frac{1}{n\mu P_n^{\mathrm{out}}},\qquad(4.23)$$

utilizing equations (4.6) and (4.10). We derive the closed form expression [HSF18a]

$$\mathrm{MTBF}_n = \frac{\exp\left(\frac{1}{\Phi}\right) - 1}{n f_{\mathrm{D}} \sqrt{\frac{2\pi}{\Phi}} \left(1 - \exp\left(\frac{-1}{\Phi}\right)\right)^n}\qquad(4.24)$$

by inserting the transition rates (3.7).

In Fig. 4.3, the MTBF$_n$ statistics are shown for selection-combined Rayleigh fading channels, normalized to the Doppler frequency f_{D}. Similar to the steady-state channel availability, the plotted MTBF$_n$ values vary over several orders of magnitude. Adding channels or increasing the fading margin Φ causes a longer MTBF$_n$ for a certain Doppler frequency f_{D}. Most importantly, however, it cannot be determined whether a high value of MTBF_n corresponds to a high MUT_n or a high MDT_n, because the MTBF_n is the sum of MDT_n and MUT_n, according to (4.19). Therefore, only investigating MTBF_n values may lead to confusion. For the considered evaluation scenario, the MTBF_n is dominated by MUT_n, which will be discussed in Section 4.1.4.

4.1.3 Mean Downtime

Definition 4. *The **mean downtime** MDT identifies the mean duration of a system failure, defined as the mean time from when the system enters a down state until it is repaired and transitions back to an up state [HR09].*

This essential dependability metric allows conclusions about a system's capability to self-repair or recover after a failure.

The complementary steady-state availability $1 - A_n$ is equal to the frequency ω_n of system failures multiplied by the MDT. Applied to the introduced scenario, the MDT from a user's viewpoint can be calculated as

$$\mathrm{MDT}_n = \frac{1 - A_n}{\omega_n}. \tag{4.25}$$

MDT of Selection-Combined Rayleigh Fading Channels

Selection combing of n Rayleigh fading links yields

$$\mathrm{MDT}_n = \frac{1}{n\mu}, \tag{4.26}$$

employing the steady-state availability expression (4.10) together with the MTBF_n (4.22). The obtained MDT_n is independent of the failure rate λ because this metric only considers downtimes, i. e., all channels are failed and a single channel repair leads to a system repair. Thus, the failures and the corresponding failure rates are irrelevant. By substituting the transition rates (3.7), we determine the closed form expression [HSF18a]

$$\mathrm{MDT}_n = \frac{\exp\left(\frac{1}{\Phi}\right) - 1}{n f_{\mathrm{D}} \sqrt{\frac{2\pi}{\Phi}}}. \tag{4.27}$$

The MDT_n shown in Fig. 4.4 decreases for larger values of Φ and a fixed f_{D}. A higher degree of redundancy, corresponding to larger n, implies a shorter MDT_n. The gaps between different levels of redundancy n remain constant, regardless of the fading margin Φ, since the MDT_n linearly depends on the reciprocal of the number of links n.

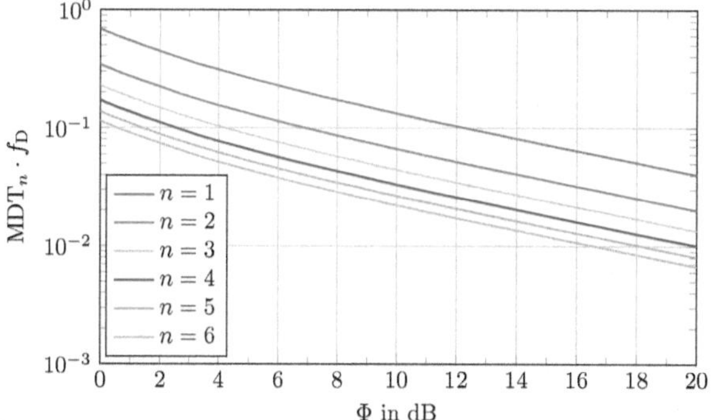

Fig. 4.4.: MDT for selection-combined Rayleigh fading channels and varying values of fading margin Φ and the number of selection-combined channels n.

MDT with Rice Fading and Interference

We extend the evaluation of the considered dependability metrics by focusing on two causes of failures, i.e., Rice fading and interference, as introduced in Section 3.3. According to (4.25), the mean downtime for a user with $n \in \{1, 2\}$ channel(s) results in [HSF19a]

$$\text{MDT}_n = \frac{1 - A^{\text{I}} A^{\text{F}}}{n A^{\text{I}} A^{\text{F}} \left(\lambda_{\text{I}} + \lambda_{\text{F}}\right)}. \tag{4.28}$$

In analogy to the previous evaluation, the MDT_n linearly scales with the reciprocal of the number n of parallel channels because if all channels are failed, any single channel repair will lead to a system repair. However, in contrast to pure small-scale fading, the MDT_n also depends on the failure rates λ_{I}, λ_{F} because the corresponding CTMC has more than one down state. Thus, there are several options to leave the set of down states. Subsequently, the following exemplary scenario is considered for further evaluating Rice fading and interference: medium velocity $v = 10\,\text{m/s}$ and the carrier frequency $f = 2\,\text{GHz}$, leading to $f_D = 66.7\,\text{Hz}$. We compare the performance for different Rice K-factors and interference outage probabilities $P^{\text{I,out}} = 1 - A^{\text{I}}$ with an interference repair rate of $\mu_{\text{I}} = 0.1\,\text{Hz}$, if not stated otherwise.

Fig. 4.5 presents the MDT_n for $n \in \{1, 2\}$ channel(s). The MDT_1 results are upper bounded by $1/\mu_{\text{I}} = 10\,\text{s}$, which corresponds to the MDT of the interference component in the series structure of a single channel. This observation is confirmed by investigating the overall MDT (4.28) for high fading margins Φ,

$$\lim_{\Phi \to \infty} \text{MDT}_n = \frac{1}{n \mu_{\text{I}}}, \tag{4.29}$$

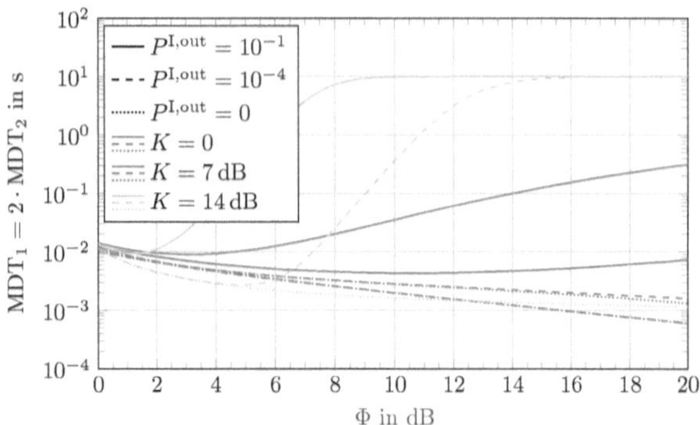

Fig. 4.5.: MDT_n for $n \in \{1, 2\}$ channel(s) depending on the fading margin Φ for different Rice K-factors and different interference outage probabilities $P^{\mathrm{I,out}}$ at Doppler frequency $f_\mathrm{D} = 66.7\,\mathrm{Hz}$ and interference repair rate $\mu_\mathrm{I} = 0.1\,\mathrm{Hz}$.

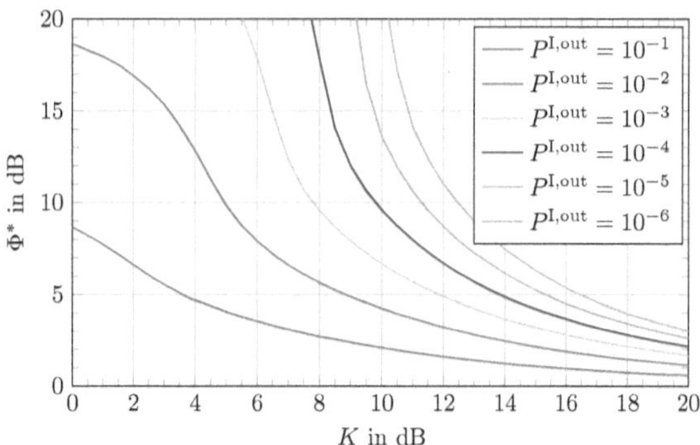

Fig. 4.6.: Optimal fading margin depending on the Rice K-factor for and different interference outage probabilities $P^{\mathrm{I,out}}$.

which corresponds to the MDT of n selection-combined channels assuming interference as the sole cause of failure, cf. (4.26). An optimum is visible for $K > 0$, minimizing the MDT, which is a conceivable objective for the system design. The optimal value does not depend on the number n of parallel channels. Fig. 4.6 shows the numerically determined optimal fading margin $\Phi^* = \arg\min_\Phi \mathrm{MDT}_n$ depending on the Rice K-factor for several interference outage probabilities $P^{\mathrm{I,out}}$. The optimal fading margin Φ^* decreases for increasing Rice K-factors and higher interference outage probabilities $P^{\mathrm{I,out}}$. The differences among the Φ^* results reduce towards larger Rice K-factors.

4.1.4 Mean Uptime

Definition 5. *The **mean uptime** MUT characterizes the mean system operational time until a failure occurs. Applied to the considered scenario, it is defined as the mean time from a transition to an up state until the first transition back to a down state [HR09].*

Utilizing the general relation (4.19) together with (4.21) and (4.25), we obtain the MUT for a user

$$\mathrm{MUT}_n = \frac{A_n}{\omega_n} = A_n \cdot \mathrm{MTBF}_n. \tag{4.30}$$

MUT of Selection-Combined Rayleigh Fading Channels

In the special case of n selection-combined Rayleigh fading links we insert equations (4.10) and (4.22) obtaining

$$\mathrm{MUT}_n = \frac{(\lambda + \mu)^n}{n \mu \lambda^n} - \frac{1}{n\mu}. \tag{4.31}$$

Applying the transition rates (3.7) yields the closed form expression [HSF18a]

$$\mathrm{MUT}_n = \frac{\exp\left(\frac{1}{\Phi}\right) - 1}{n f_\mathrm{D} \sqrt{\frac{2\pi}{\Phi}}} \left(\frac{1}{\left(1 - \exp\left(\frac{-1}{\Phi}\right)\right)^n} - 1 \right). \tag{4.32}$$

It turns out that, in contrast to the steady-state channel availability A_n, the user's MTBF_n, MDT_n, and MUT_n depend on the fading margin Φ and maximum Doppler shift f_D. Hence, we propose to utilize these KPIs for the research on wireless communications systems, because these metrics enable to evaluate the dependability of the wireless communications system from the user's viewpoint taking into account the actual failure and repair rates λ and μ. It is obvious that system dependability is based upon the quantities MUT_n, MDT_n besides channel availability A_n, which can be linked by the following relation

$$A_n = \frac{\mathrm{MUT}_n}{\mathrm{MTBF}_n} = \frac{\mathrm{MUT}_n}{\mathrm{MUT}_n + \mathrm{MDT}_n}. \tag{4.33}$$

As shown in Fig. 4.7, the MUT_n is higher for larger numbers of channels n and for higher values of the fading margin Φ. The differences between the curves increase for larger values of the fading margin Φ for a given Doppler frequency f_D. At high fading margins, adding a channel can improve the normalized MUT_n by more than one order of magnitude. After analyzing the different metrics individually, we subsequently study them jointly. Several cases with different values for the carrier frequency f and velocity v may all lead to the same steady-state channel availability A_n. In Table 4.1, we evaluate combinations of low and high velocity v with different carrier frequencies f for

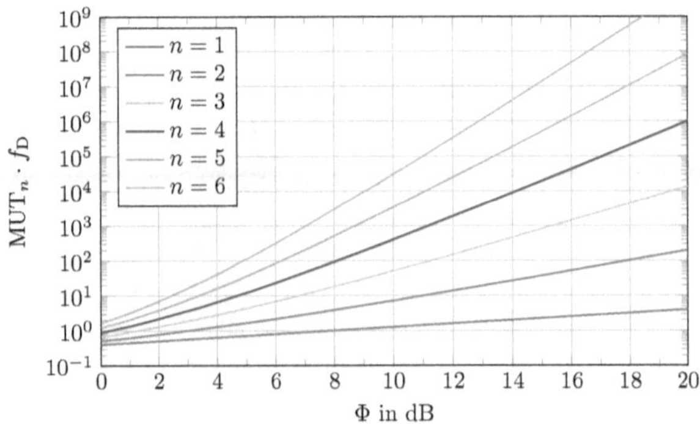

Fig. 4.7.: MUT_n for selection-combined Rayleigh fading links and varying values of fading margin Φ and number of channels n.

Tab. 4.1.: Exemplary Comparison of MUT_n, MDT_n and complementary channel availability of n selection-combined Rayleigh fading channels for fading margin $\Phi = 20\,\mathrm{dB}$

n	P_n^{out}	v [m/s]	f [GHz]	MDT_n	MUT_n
3	10^{-6}	10	2	0.2 ms	3.4 min
3	10^{-6}	10	60	6.7 µs	6.8 s
3	10^{-6}	80	2	25.0 µs	25.4 s
3	10^{-6}	80	60	0.8 µs	0.8 s
5	10^{-10}	10	2	0.1 ms	14.3 d
5	10^{-10}	10	60	4.0 µs	11.4 h
5	10^{-10}	80	2	15.0 µs	42.8 h
5	10^{-10}	80	60	0.5 µs	1.4 h

$n = 3$ and $n = 5$ redundant links, confining our concentration on the exemplary fading margin $\Phi = 20\,\mathrm{dB}$. The probability of outage decreases by two orders of magnitude per additional channel, e. g., $n = 3$ selection-combined channels already yield a channel availability of $A_3 = 99.9999\,\%$. Obviously, multiple system designs with the same outage probability P_n^{out} can exhibit significantly varying MUT_n and MDT_n. For $n = 3$ links, the obtained outage probability $P_n^{\mathrm{out}} = 10^{-6}$ appears promising for many URLLC use cases, but the MUT_3 differs by orders of magnitude in the range between 800 ms and 3.4 min. The corresponding MDT_3 values are comparable to strict latency requirements for URLLC applications, e. g., wireless factory automation. Thus, it is obvious that the reliability requirements cannot be satisfied permanently if the MDT is in the range of the latency constraints, even though current systems should be able to tolerate short downtimes. As expected, two additional redundant links improve the outage probability P_5^{out} by a factor of 10 000. The MUT_5 also increases significantly, the expected uptime of more than two weeks for the low mobility and low frequency scenario could be similar to the maintenance cycle in factory automation. However, the MDT_5 is not even reduced by half.

MUT with Rice Fading and Interference

Considering two causes of failures, i. e., Rice fading and interference, yields [HSF19a]

$$\mathrm{MUT}_1 = \frac{1}{\lambda_\mathrm{I} + \lambda_\mathrm{F}}, \tag{4.34a}$$

$$\mathrm{MUT}_2 = \frac{2 - A^\mathrm{I} A^\mathrm{F}}{2\left(1 - A^\mathrm{I} A^\mathrm{F}\right)\left(\lambda_\mathrm{I} + \lambda_\mathrm{F}\right)} \tag{4.34b}$$

as the mean uptime for a user with $k \in \{1, 2\}$ channel(s), respectively. In contrast to the metric availability A_n, the user's MDT_n and MUT_n depend on the fading margin Φ, Rice K-factor, and maximum Doppler shift f_D, included in λ_F, as well as the interference failure rate λ_I and the interference availability A^I. Hence, we propose to utilize these KPIs for the research on wireless communications systems, because these metrics allow to evaluate the dependability of a wireless communications system from the user's viewpoint taking into account all actual rates λ_I, λ_F, μ_I, and μ_F. The special case of Rayleigh fading ($K = 0$) without interference ($\lambda_\mathrm{I} = 0$, $A_\mathrm{I} = 1$) leads to the dependability metric expressions determined above.

Fig. 4.8 illustrates the MUT_n for $n \in \{1, 2\}$ channel(s), respectively. As expected, the value range is significantly larger compared to the MDT_n in Fig. 4.5. The MUT_n increases for larger fading margin Φ, Rice K-factor, number k of parallel channels, and lower interference outage probability $P^{\mathrm{I,out}}$. The highest gain is visible for large fading margins Φ: For instance, comparing $n = 1$ Rayleigh-faded ($K = 0$) channel with $n = 2$ parallel channels exhibiting a clear LOS (Rice fading with $K = 14\,\mathrm{dB}$) leads to an improvement from $\mathrm{MUT}_1 = 100\,\mathrm{ms}$ to $\mathrm{MUT}_2 > 15$ years at $\Phi = 20\,\mathrm{dB}$ and $P^{\mathrm{I,out}} = 10^{-4}$. In analogy to the observations above, the overall MUT is bounded due to the interference component for higher fading margins Φ, which can be derived from (4.34) according to

$$\lim_{\Phi \to \infty} \mathrm{MUT}_1 = \frac{1}{\lambda_\mathrm{I}}, \tag{4.35a}$$

$$\lim_{\Phi \to \infty} \mathrm{MUT}_2 = \frac{2\lambda_\mathrm{I} + \mu_\mathrm{I}}{2\lambda_\mathrm{I}^2}. \tag{4.35b}$$

These values correspond to the MUT_n of $n \in \{1, 2\}$ selection-combined channels with interference as the single cause of failure, cf. (4.31).

These evaluations demonstrate that it is of key importance to consider the metrics MUT_n, MDT_n, and MTBF_n leading to a joint analysis of availability and time-related dependability metrics also in wireless communications. In contrast to the steady-state channel availability A_n and outage probability P_n^{out}, these quantities depend on the maximum Doppler shift f_D, reflecting the impact of the carrier frequency and mobility aspects.

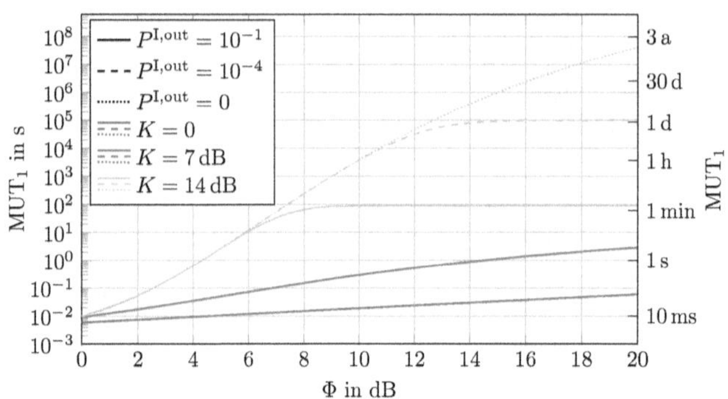

(a) Single channel.

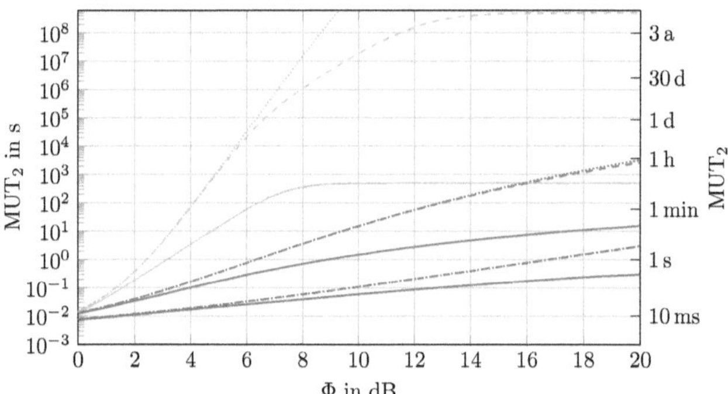

(b) Two selection-combined channels.

Fig. 4.8.: MUT_n for $n \in \{1, 2\}$ depending on the fading margin Φ with different Rice K-factors and different interference outage probabilities $P^{\mathrm{I,out}}$ at Doppler frequency $f_\mathrm{D} = 66.7\,\mathrm{Hz}$. The legend in (a) also applies for (b).

The dependability metrics MUT_n, MDT_n, and MTBF_n can only give an estimate whether a system supports particular use cases or not, since they are mean quantities. Aiming for concrete performance guarantees requires to study the detailed distributions of uptimes and downtimes, which is presented in this section for selection-combined Rayleigh fading channels.

Downtime Distribution

Definition 6. *The **downtime** T^{d} specifies the duration from a transition to a down state until the first transition back to an up state.*

According to the system model for selection-combined Rayleigh fading channels introduced in Chapter 3, the CTMC comprises a single down state $\mathcal{U} = \{0\}$. Hence, the downtime of the system corresponds to the sojourn time in state 0. The expectation of T^{d} constitutes the MDT_n, derived in (4.26) and (4.27), according to

$$E\left[T^{\mathrm{d}}\right] = \mathrm{MDT}_n = \frac{1}{n\mu} = \frac{\exp\left(\frac{1}{\Phi}\right) - 1}{n f_{\mathrm{D}} \sqrt{\frac{2\pi}{\Phi}}}. \tag{4.36}$$

Due to the assumed Markov property, i.e., the future development of the stochastic process only depends on the current state and is independent of anything that has happened in the past, the downtime is exponentially distributed,

$$T^{\mathrm{d}} \sim \mathrm{Exp}\left(\frac{1}{\mathrm{MDT}_n}\right). \tag{4.37}$$

In [ÖF15], this approximation of the Rayleigh fade duration distribution by an exponential distribution has been evaluated by comparison to simulations and preexisting work, revealing high accuracy. The cumulative distribution function (CDF) of the downtime T^{d} results in the explicit expression [Höß+19]

$$G_n\left(t^{\mathrm{d}}\right) = 1 - \exp\left(-n\mu t^{\mathrm{d}}\right) = 1 - \exp\left(\frac{-\sqrt{\frac{2\pi}{\Phi}} n f_{\mathrm{D}} t^{\mathrm{d}}}{\exp\frac{1}{\Phi} - 1}\right). \tag{4.38}$$

Fig. 4.9 illustrates the complementary cumulative distribution function (CCDF) $1 - G_n$ of downtimes for the considered scenario. Of course, shorter downtimes are aimed for in URLLC applications. Higher numbers of channels reduce the downtime for a fixed percentile. For any percentile, the ratio between the downtime t_1^{d} with n_1 channels and t_2^{d} with n_2 channels is the constant n_2/n_1 due to the exponential distribution. Thus, the gain for adding channels decreases with each additional channel. The probability of

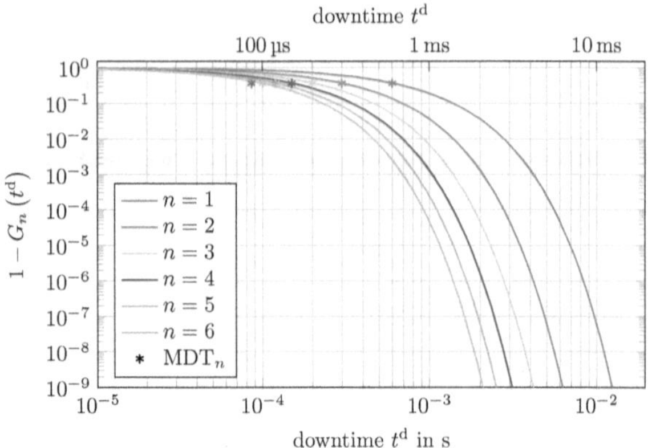

Fig. 4.9.: Downtime CCDF for Doppler frequency $f_{\mathrm{D}} = 66.7\,\mathrm{Hz}$, fading margin $\Phi = 20\,\mathrm{dB}$, and different numbers of channels n.

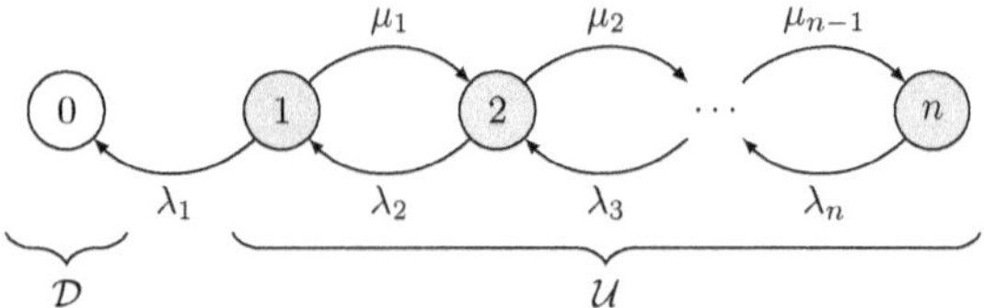

Fig. 4.10.: Selection combining of n channels as modified CTMC with an absorbing down state.

exceeding the MDT_n is about $36.8\,\%$. Thus, the MDT_n is of limited benefit for URLLC. A reasonable requirement in the context of URLLC could be to guarantee a maximum tolerated downtime with a certain probability: for instance, a downtime threshold of $1\,\mathrm{ms}$ is kept by a single channel with a probability of $20\,\%$. The probability of not exceeding $1\,\mathrm{ms}$ downtime can be increased to $99\,\%$ or more than $99.99\,\%$ by raising the number of channels n to 3 or 6, respectively.

Uptime Distribution

Definition 7. *The **uptime** T^{u} is defined as the duration from a transition to an up state until the first transition back to a down state.*

In order to determine the uptime distribution, the introduced CTMC with states $0, 1, \ldots, n$ is accordingly modified by setting $\mu_0 = 0$, shown in Fig. 4.10: The down state 0 is a single absorbing state and the process is considered to start in state 1, which is the state of arrival after the transition from the down state.

The modified CTMC has the transition rate matrix

$$\hat{M}_{\mathrm{T}} = \begin{pmatrix} 0 & \mathbf{0} \\ \boldsymbol{S}_0 & S \end{pmatrix}, \tag{4.39}$$

where the $n \times n$ matrix S includes the transition rates among the up states. The absorption rates are contained in the vector

$$\boldsymbol{S}_0 = -S\mathbf{1} \tag{4.40}$$

with $\mathbf{1}$ describing a $n \times 1$ vector, where every element is equal to 1. Please note that the modified transition matrix $\hat{M}$ only differs from the original matrix M (3.3) in the first row.

A probability distribution with CDF $F(\cdot)$ on $[0, \infty)$ is of phase-type if it can arise as the absorption time distribution of an CTMC with n states and the absorbing state 0 [Neu94]. The non-absorbing states are referred to as phases. Thus, the uptime T^{u}, considered in this work, follows a phase-type distribution, denoted as $T^{\mathrm{u}} \sim \mathrm{PH}(\boldsymbol{\alpha}, S)$. The CDF of the uptime is given by [Höß+19]

$$F(t^{\mathrm{u}}) = 1 - \boldsymbol{\alpha} \exp(St^{\mathrm{u}})\mathbf{1}, \tag{4.41}$$

where $\exp(\cdot)$ denotes the matrix exponential and the vector $\boldsymbol{\alpha}$ contains the initial probabilities of the up states [Neu94]. Since the modified CTMC for the considered wireless system is assumed to start in state 1, the initial probabilities of the up states are $\boldsymbol{\alpha} = (1, 0, 0, \ldots, 0)$. The expectation of T^{u} corresponds to the mean uptime (MUT), resulting in

$$E[T^{\mathrm{u}}] = \mathrm{MUT}_n = -\boldsymbol{\alpha}S^{-1}\mathbf{1}. \tag{4.42}$$

Incorporating the rates (3.7) confirms the expressions (4.31) and (4.32) presented in the previous section.

The considered diversity system can be approximated by a single unit with the constant failure rate

$$\tilde{\lambda} = \frac{1}{\mathrm{MUT}_n} = \frac{1}{-\boldsymbol{\alpha}S^{-1}\mathbf{1}}. \tag{4.43}$$

This corresponds to approximating the phase-type distributed uptimes by an exponential distribution

$$F(t^{\mathrm{u}}) \approx \tilde{F}(t^{\mathrm{u}}) = 1 - \exp\left(\frac{-t^{\mathrm{u}}}{-\boldsymbol{\alpha}S^{-1}\mathbf{1}}\right), \tag{4.44}$$

which will be utilized in Section 4.2.2.

In the following, the uptime distribution is evaluated with respect to the exemplary scenario introduced above, comprising velocity $v = 10\,\text{m/s}$ and the carrier frequency $f = 2\,\text{GHz}$, leading to the Doppler frequency $f_\text{D} = 66.7\,\text{Hz}$. Again, the index n relates to the number of Rayleigh fading channels a user is simultaneously connected to, performing selection combining. A fading margin of $\Phi = 20\,\text{dB}$ is assumed, which leads to a channel availability $A_1 = 99\,\%$ for a single channel according to (4.10).

In Fig. 4.11a we compare the phase-type distributed uptime CDF F_n with the proposed approximation $\tilde{F}_n$ for different numbers of channels n. The probability of not achieving the MUT_n is about $63.2\,\%$. Thus, we propose to refine the discussion on performance guarantees in the context of URLLC by utilizing percentiles of the uptime CDF. For instance, if a continuous uptime of $1\,\text{s}$ should be attained with $99\,\%$ probability, at least $n = 3$ selection-combined channels are necessary. Moreover, it turns out that even high numbers of channels n are unable to guarantee relatively long uptimes in the considered scenario with substantially higher probability, such as $99.9\,\%$. The reason is that the phase-type distributions of uptimes approach F_1 for small values of t^u. On the other hand, the approximations $\tilde{F}_n$ approach the phase-type distributions F_n for high uptimes. Fig. 4.11b depicts the corresponding relative approximation error

$$d_n(t^\text{u}) = \frac{\left| F_n\left(t^\text{u}\right) - \tilde{F}_n\left(t^\text{u}\right) \right|}{F_n\left(t^\text{u}\right)}. \tag{4.45}$$

It can be observed that the approximation error vanishes for large uptimes. The notches show that the approximation crosses the exact distribution. The consideration of a single channel, $n = 1$, is the special case of a phase-type distribution with one phase, equivalent to an exponential distribution. Thus, the corresponding approximation error is zero, $d_1(t^\text{u}) = 0$.

A special case of the metric uptime is the time to first failure, defined as follows.

Definition 8. *The **time to first failure** is the time elapsing from when an item is put into operation until it fails for the first time.*

This KPI originates from dependability theory of non-repairable technical components, e. g., electronic devices, which are assumed to be free from defects at the time of commissioning [HR09]. The condition of no failures at the beginning of the observation can be adopted to the introduced system model via the following strategy: The communications system, consisting of selection-combined Rayleigh fading channels, is started when all n channels are operational. Thus, the CTMC has the initial state n, corresponding to

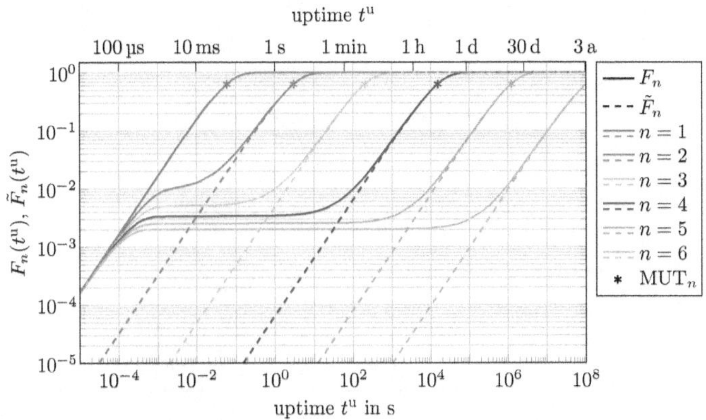

(a) Uptime CDF (solid) and approximation (dashed).

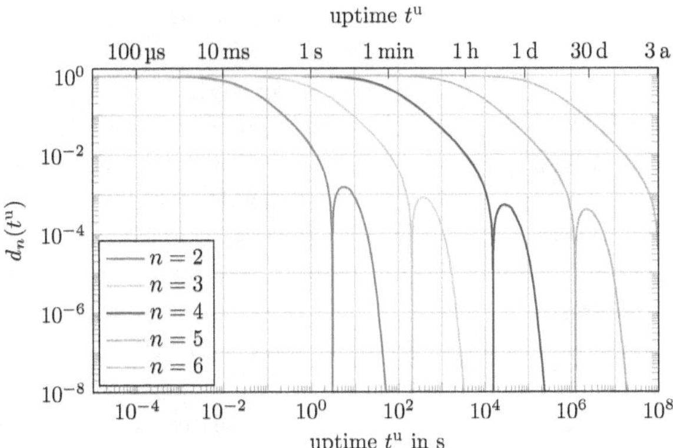

(b) Error of uptime CDF approximation.

Fig. 4.11.: Uptime CDF, approximation, and error for Doppler frequency $f_D = 66.7\,\text{Hz}$, fading margin $\Phi = 20\,\text{dB}$, and different numbers of channels n.

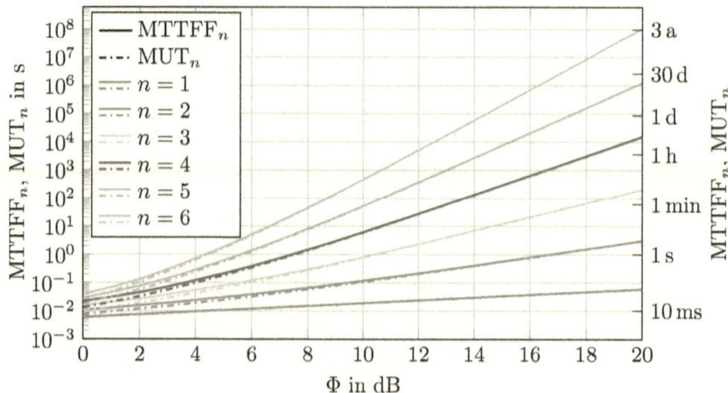

Fig. 4.12.: Comparison of MTTFF$_n$ and MUT$_n$ at Doppler frequency $f_\mathrm{D} = 66.7\,\mathrm{Hz}$.

the initial state probability vector $\boldsymbol{\alpha}_\mathrm{FF} = (0, 0, \ldots, 0, 1)$. Applying (4.42) with $\boldsymbol{\alpha} = \boldsymbol{\alpha}_\mathrm{FF}$ leads to the analytic expression of the *MTTFF*, i. e.,

$$
\mathrm{MTTFF}_n = \frac{1}{\sqrt{\frac{2\pi}{\Phi}}\,f_\mathrm{D}} \sum_{j=1}^{n} \frac{\sum_{m=1}^{j} \left(\exp\frac{1}{\Phi} - 1\right)^{m-1} (m-1)!\,\prod_{\ell=m}^{j-1} (n-\ell)}{\left(\exp\frac{1}{\Phi} - 1\right)^{j-1} j!},
\tag{4.46}
$$

which was first derived and evaluated for the considered system model in our publication [HSF18b]. Similar to the MUT$_n$, the MTTFF$_n$ takes the carrier frequency and mobility into account, because it depends on the fading margin Φ and the maximum Doppler frequency f_D. The relationship between the MTTFF$_n$ and the Doppler frequency f_D remains inversely proportional, independent of the number of channels. In Fig. 4.12 the MTTFF$_n$ is compared with the MUT$_n$ at Doppler frequency $f_\mathrm{D} = 66.7\,\mathrm{Hz}$ for different numbers of channels n and fading margins Φ. The curves are very similar and the differences decrease for larger fading margins. As expected, the MTTFF$_n$ is an upper bound for the MUT$_n$ due to the different initial states. In order to determine the MTTFF$_n$, the CTMC starts in state n, which is the best case because all up states need to be passed until absorption in the down state. On the other hand, the MUT$_n$ is derived starting from the first transition from the down state into the up states, i. e., state 1, which is the worst case because down state is closest. In the trivial case of $n = 1$ channel, the MTTFF$_1$ and the MUT$_1$ coincide due to the single up state of the CTMC.

4.1.6 Application Availability

Some applications can bridge short communication outages with appropriate design, e. g., fault-tolerant controllers. Thus, an application outage event occurs only when the communications system exhibits a down state which lasts for more than a tolerable threshold. In the context of this work, these applications are called robust and the following novel KPI definition is proposed, which covers this property:

Definition 9. *Application availability* A^A *is defined as the probability of not experiencing any downtimes which are longer than the maximum tolerated downtime* t^d_{max}. *The complementary application availability is denoted as* **application outage probability** $P^{A,out}$, *i. e.,*

$$A^A\left(t^d_{max}\right) = 1 - P^{A,out}\left(t^d_{max}\right) = 1 - \Pr\left[p < p_{min} \wedge t^d > t^d_{max}\right]. \tag{4.47}$$

For a sufficiently long observation duration T, the application outage probability can be expressed by the fraction of time the system experiences outages that are longer than the threshold t^d_{max}, i. e.,

$$P^{A,out} = \frac{\sum_{t^d \in \mathcal{C}(T)} t^d}{T}, \tag{4.48}$$

where the set of downtimes that cause application outages is denoted by

$$\mathcal{C}(T) = \left\{t^d \mid t^d \in \mathcal{T}(T),\ t^d > t^d_{max}\right\} \tag{4.49}$$

and the set $\mathcal{T}(T)$ contains every individual downtime t^d during the time interval $[0, T]$. The KPI channel availability, according to Definition 2 introduced in Section 4.1.1, is a special case of the application availability, where the maximum tolerated downtime is zero, i. e.,

$$A = A^A(0). \tag{4.50}$$

Application outage probability, as defined above in Definition 9, translates to the concept of minimum duration outage, published in [MCH96], where the focus is on shadow fading. However, the mathematical definition proposed above is more precise. Furthermore, by utilizing the term application availability, we emphasize the application's ability to tolerate short outage times. In [ÖF15], an analytical expression of this KPI has been derived and evaluated for multiple selection-combined Rayleigh fading links. Their fade duration distributions are assumed to be exponential, which complies with the system model introduced in Chapter 3. The numerical evaluation in [ÖF15] revealed high accuracy of the exponential approximation with respect to simulation independent of the fading margin used. In Chapter 5, the metric application availability will be applied to evaluate different dynamic connectivity approaches for robust applications.

4.2 Mission-Related Dependability KPIs

The metrics previously discussed characterize dependability aspects of a system, referring
to a time instant of observation. They are unable to analyze individual time intervals,
which are, however, of major interest for URLLC because specifying performance guarantees of continuing a failure-free operation throughout a time interval is required, e. g.,
during maneuvers of wirelessly controlled robots or self-driving cars. We denote this
time interval as "mission" of duration Δt, which is the key element of the contributions
presented in this section: The dependability quantities *mission reliability* and *mission
availability* will be introduced, characterizing the success probability of experiencing no
or only short failures during a certain mission duration Δt, respectively. Both metrics
will be evaluated regarding the considered system model comprising selection-combined
Rayleigh fading channels. Trade-offs between mission availability or mission reliability,
mission duration, number of channels, and tolerated downtime will be demonstrated for
an exemplary scenario.

4.2.1 Mission Reliability

The wireless communications community interprets reliability as percentage of successful
packet delivery [3GP17c]. In Section 4.1.1 we have shown that by applying fundamental
dependability theory, it is evident that this metric falls under the definition of availability
(cf. Definition 2) as opposed to reliability. In this section, we consider the reliability
interpretation as it is common in dependability theory:

Definition 10. *Mission reliability* $R(\Delta t)$ *is the probability that an item is able to perform
as required throughout a mission time interval Δt.*

The mission reliability can be expressed according to

$$R(\Delta t) = \Pr\left\{x(\tau) = 1 \,\forall\, \tau \in [0, \Delta t]\right\} \tag{4.51}$$

with the binary, time-dependent state variable

$$x(\tau) = \begin{cases} 0, & \text{if the item is in a failed state ("down") at time } \tau, \\ 1, & \text{if the item is in an operational state ("up") at time } \tau. \end{cases} \tag{4.52}$$

In order to emphasize the failure-free operation during this time interval, the traditional
reliability definition (e. g., in [BP65; ITU08]) is extended by the term "mission" here,
because it is important to understand that referring to a certain reliability value without
specifying the corresponding mission duration Δt is not a valid statement. In general, it
is not possible to convert between mission reliability and channel availability because

mission reliability depends on the mission duration as opposed to solely channel availability. In wireless systems, however, reliability is usually defined as the amount (in %) of sent packets successfully delivered to a given node divided by the total number of sent packets [3GP16a]. This interpretation is related to PLR or outage probability without reference to the time dimension. Since a failure-free operation is practically impossible for long missions due to the random processes causing failures, the limiting value of the mission reliability $R(\Delta t)$ as the mission duration Δt approaches infinity is always zero,

$$\lim_{\Delta t \to \infty} R(\Delta t) = 0. \tag{4.53}$$

This is fundamentally different to the instantaneous channel availability, which converges to the steady-state channel availability, according to (4.2).

In order to determine the mission reliability of the considered wireless communications system, the modified CTMC from Section 4.1.5 is utilized again, assuming failed states to be absorbing. The differential balance equations retain the same structure as in (3.1). The corresponding transition rates in (3.2) also remain the same, except for $\mu_0 = 0$. The state probabilities $\hat{P}_j(\Delta t)$ of the modified Markov model denote the probability that j channels are operational at the end of mission duration Δt and that the number of operational channels has never reached zero during the mission duration Δt. The mission reliability [HSF18b]

$$R_n(\Delta t) = \sum_{j \in \mathcal{U}} \hat{P}_j(\Delta t) = \sum_{j=1}^{n} \hat{P}_j(\Delta t) \tag{4.54}$$

characterizes the probability of not leaving the set of operational states $\mathcal{U}$ during the mission duration Δt for a system with n redundant components, e. g., selection-combined Rayleigh-fading channels. In the considered system model, the maximum Doppler frequency f_D purely acts as a scaling factor to the mission duration Δt, since the maximum Doppler frequency f_D is a factor in both the transition rates (3.7) and the balance equations (3.1).

The complexity of deriving the mission reliability increases with the number of channels, because it requires solving the differential balance equation system

$$\dot{\hat{P}}_j(t) = \mu_{j-1}\hat{P}_{j-1}(t) - (\lambda_j + \mu_j)\hat{P}_j(t) + \lambda_{j+1}\hat{P}_{j+1}(t) \tag{4.55}$$

for $j = 1, \ldots, n$ with $\hat{P}_j(0) = 0$ for $j \neq n$. As discussed in Section 4.1.5, it is reasonable to assume all n channels to be operational at time $t = 0$, equivalent to $\hat{P}_n(0) = 1$. The mission reliability of the single-connectivity, i. e., $n = 1$, results in

$$R_1(\Delta t) = \exp\left(\frac{-\Delta t}{\text{MTTFF}_1}\right) \tag{4.56}$$

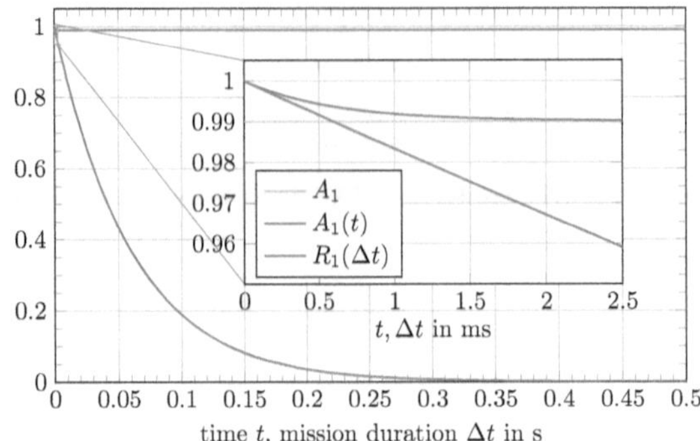

Fig. 4.13.: Instantaneous channel availability, steady-state channel availability, and mission reliability for a single Rayleigh-fading link, Doppler frequency $f_\mathrm{D} = 66.7\,\mathrm{Hz}$, and fading margin $\Phi = 20\,\mathrm{dB}$.

with

$$\mathrm{MTTFF}_1 = \frac{1}{\lambda} = \frac{1}{\sqrt{\frac{2\pi}{\Phi}} f_\mathrm{D}}. \tag{4.57}$$

This complies with the fact that the time to first failure of a single item with constant failure rate λ is exponentially distributed [HR09]. Based on this result, we model the considered diversity system as a single item and propose the approximation [HSF18b]

$$\tilde{R}_n(\Delta t) = \exp\left(\frac{-\Delta t}{\mathrm{MTTFF}_n}\right) \tag{4.58}$$

with the constant failure rate $1/\mathrm{MTTFF}_n$ determined from the closed form expression (4.46). The accuracy of this approximation is discussed at the end of this section.

Subsequently, we evaluate the proposed performance metrics with respect to the exemplary scenario comprising a typical fading margin $\Phi = 20\,\mathrm{dB}$, medium velocity $v = 10\,\mathrm{m/s}$, and the carrier frequency $f = 2\,\mathrm{GHz}$, equivalent to $f_\mathrm{D} = 66.7\,\mathrm{Hz}$, if not stated otherwise. All n channels are assumed to be operational at time $t = 0$.

In Fig. 4.13, we focus on the special case of single-connectivity, $n = 1$, presenting fundamental differences between (instantaneous) channel availability and mission reliability. The instantaneous channel availability $A_1(t)$ converges quickly to the steady-state channel availability $A_1 = 0.99$. Thus, in the steady state, an outage probability of $1\,\%$ is expected at any instant of time. On the other hand, the mission reliability $R_1(\Delta t)$ approaches zero, because a failure is sure to occur sooner or later due to the random fading process. Hence, in the context of URLLC with different mission duration it is necessary to analyze and improve the mission reliability of wireless communication channels, e. g., by means of multi-connectivity.

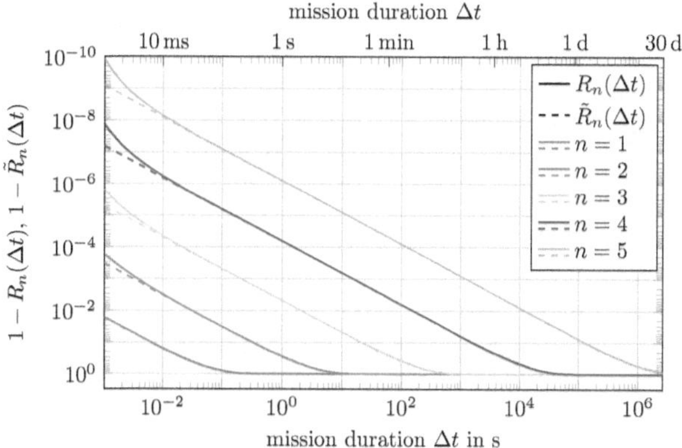

Fig. 4.14.: Complementary mission reliability and corresponding approximation for n selection combined Rayleigh-fading links, Doppler frequency $f_\mathrm{D} = 66.7\,\mathrm{Hz}$, and fading margin $\Phi = 20\,\mathrm{dB}$.

In Fig. 4.14, the impact of selection combining on the KPI mission reliability is evaluated. For readability reasons, the complementary mission reliability is depicted on a reversed axis. This presentation takes advantage of the logarithmic scale in the relevant range, emphasizing that values on top are superior, which is common in dependability theory. The trade-off between mission duration, mission reliability, and number of links is revealed, which can be applied to specify requirements for URLLC: In contrast to the previously discussed KPIs, we can define the performance threshold by a target mission duration and the corresponding mission reliability. For instance, if a mission reliability of more than $99.999\,\%$ is required for an autonomous car throughout the mission of $10\,\mathrm{s}$ crossing an intersection, at least $n = 5$ selection-combined links are necessary. In addition, the accuracy of the proposed approximation (4.58) can be analyzed by comparing it with the mission reliability resulting from equation (4.54). It is clearly visible that the approximation $\tilde{R}_n$ is a lower bound for the mission reliability R_n and for mission duration $\Delta t > 10\,\mathrm{ms}$ the approximation shows a very good match with respect to the considered scenario.

4.2.2 Mission Availability

The previous section on mission reliability demonstrated how the success probability of a continuous mission, which does not allow any failure, depends on its mission duration. However, as introduced in Section 4.1.6, many of today's applications are robust, i. e., they can tolerate short communication outage times due to appropriate mechanisms, e. g., controllers. In the following, the success probability of a mission is analyzed, assuming that the system is able to tolerate short interruptions. In this context, application availability, as defined in Section 4.1.6, is a related quantity, characterizing the probability of experiencing failures if the power level remains below a threshold for a longer duration than a tolerated maximum downtime. Unfortunately, this metric is an average probability which does not consider the actual mission duration. Instead, the following generalization of the KPI mission reliability covers this aspect, which has been introduced in dependability theory by [Bir85], but it has not yet been utilized in wireless communications:

Definition 11. *Mission availability* $M\left(\Delta t, t_{\max}^{\mathrm{d}}\right)$ *is defined as the probability that all downtimes are not longer than the threshold $t_{\max}^{\mathrm{d}}$ during a mission of duration Δt.*

Thus, we propose to express mission availability by [Höß+19]

$$M\left(\Delta t, t_{\max}^{\mathrm{d}}\right) = \Pr\left[\forall t^{\mathrm{d}} \in \mathcal{T}(\Delta t) : t^{\mathrm{d}} \le t_{\max}^{\mathrm{d}}\right], \tag{4.59}$$

with the set $\mathcal{T}(\Delta t)$ containing every individual downtime t^{d} which occurs in the mission duration interval $[0, \Delta t]$. In order to determine the mission availability, all cases with $m = 0, 1, \dots$ failures have to be considered, accounting for the fact that at the end of the mission the application is operating to reach the mission duration Δt. Hence, [Bir13] sets

$$M\left(\Delta t, t_{\max}^{\mathrm{d}}\right) = 1 - F(\Delta t) + \sum_{m=1}^{\infty} C_m(\Delta t)\left(G\left(t_{\max}^{\mathrm{d}}\right)\right)^m, \tag{4.60}$$

where $(G(t_{\max}^{\mathrm{d}}))^m$ specifies the probability that all m downtimes are not longer than $t_{\max}^{\mathrm{d}}$ and $C_m(\Delta t)$ is the probability for exactly m failures during the mission duration Δt. For practical systems, $C_m(\Delta t)$ can be approximated as [Bir13]

$$C_m(\Delta t) \approx H_m(\Delta t) = U_m(\Delta t) - U_{m+1}(\Delta t) \tag{4.61}$$

with the CDF of m aggregated uptimes

$$U_m(\Delta t) = \Pr\left[\sum_{i=0}^{m-1} T_i^{\mathrm{u}} \le \Delta t\right]. \tag{4.62}$$

This metric does not take the failure duration after each uptime into consideration, which is reasonable in the context of mission availability for the following reasons: Firstly, in practical systems, the downtimes are shorter than uptimes by several orders of magnitude. Secondly, by neglecting the downtimes, the approximation (4.61) leads to a tight pessimistic bound for the mission availability (4.60), because the true sum of uptimes is shorter than the mission duration Δt. So the actual number of failures is expected to be slightly less.

The CDF of aggregated uptimes is determined by recursive convolution, according to

$$U_1(\Delta t) = F(\Delta t), \tag{4.63a}$$

$$U_{m+1}(\Delta t) = \int_0^{\Delta t} U_m(\Delta t - t) f(t) \, \mathrm{d}t \tag{4.63b}$$

with $f(t) = \mathrm{d}F(t)/\mathrm{d}t$ [Bir13]. Due to the summation of m-fold convolved functions in (4.60) with $m \to \infty$, the computation is very complex and, thus, unfeasible in the case of the phase-type distributed uptime duration. Following our proposal in (4.43), the considered diversity system can be approximated by a single unit with the constant failure rate

$$\tilde{\lambda} = \frac{1}{\mathrm{MUT}_n}. \tag{4.64}$$

This corresponds to approximating the phase-type distributed uptimes by the exponential distribution

$$F\left(t^{\mathrm{u}}\right) \approx \tilde{F}\left(t^{\mathrm{u}}\right) = 1 - \exp\left(-\tilde{\lambda}t^{\mathrm{u}}\right). \tag{4.65}$$

For a constant failure rate $\tilde{\lambda}$, the number of failures occurring during the mission duration Δt follows a Poisson distribution, i. e.,

$$H_m(\Delta t) \approx \tilde{H}_m(\Delta t) = \frac{(\tilde{\lambda}\Delta t)^m}{m!} \exp\left(-\tilde{\lambda}\Delta t\right). \tag{4.66}$$

Inserting this into (4.60) yields the approximated mission availability [Höß+19]

$$\tilde{M}\left(\Delta t, t_{\mathrm{max}}^{\mathrm{d}}\right) = \exp\left[-\tilde{\lambda}\Delta t\left(1 - G\left(t_{\mathrm{max}}^{\mathrm{d}}\right)\right)\right] = \exp\left[-\tilde{\lambda}\Delta t \exp\left(-n\mu t_{\mathrm{max}}^{\mathrm{d}}\right)\right] \tag{4.67}$$

with the CDF of the downtime (4.38). The sum term disappeared by utilizing the series representation of the exponential function

$$\sum_{m=0}^{\infty} \frac{y^m}{y!} = \exp(y). \tag{4.68}$$

In accordance to mission reliability, mission availability converges to zero for long missions,

$$\lim_{\Delta t \to \infty} M\left(\Delta t, t_{\mathrm{max}}^{\mathrm{d}}\right) = 0, \tag{4.69}$$

since there is a non-zero probability that the tolerated maximum downtime is eventually exceeded in the presence of random fading. The proof of this fundamental property can be sketched as follows: For $\Delta t \to \infty$ in (4.60), it holds that $F(\Delta t) \to 1$. Further, using (4.61) and abbreviating $G := G(t_{\max}^{\mathrm{d}})$, the sum in (4.60) tends to zero, because it can be written as:

$$U_1(\Delta t)G + \sum_{m=2}^{\infty} U_m(\Delta t)\left(G^m - G^{m-1}\right) = U_1(\Delta t)G + \left(1 - \frac{1}{G}\right)\sum_{m=2}^{\infty} U_m(\Delta t)G^m \tag{4.70a}$$

$$\xrightarrow{\Delta t \to \infty} \quad G + \frac{G-1}{G}\sum_{m=2}^{\infty} G^m = G + \frac{(G-1)G}{1-G} = 0. \tag{4.70b}$$

To obtain (4.70b) from (4.70a), the dominated convergence theorem for series was applied to move the limit over Δt into the sum. This is possible, because the series $U_m(\Delta t)G^m$ is dominated by the geometric series G^m and $U_m(\Delta t) < 1$, $G < 1$.

Obviously, mission reliability is a special case of mission availability with $t_{\max}^{\mathrm{d}} = 0$, given by

$$R(\Delta t) = M(\Delta t, 0). \tag{4.71}$$

By inserting the uptime CDF (4.41), we derive a closed-form expression of the mission reliability for the considered wireless diversity system according to [Höß+19]

$$R(\Delta t) = \boldsymbol{\alpha}_{\mathrm{FF}} \exp(S\Delta t)\mathbf{1}. \tag{4.72}$$

Consistent with Section 4.2.1, the initial state probability vector $\boldsymbol{\alpha}_{\mathrm{FF}} = (0, 0, \ldots, 0, 1)$ reflects the assumption that all n channels are operational at the beginning of the observation.

Numerical results of the proposed mission availability approximation are depicted in Fig. 4.15 for the considered exemplary multi-connectivity scenario of velocity $v = 10\,\mathrm{m/s}$ and carrier frequency $f = 2\,\mathrm{GHz}$, leading to the Doppler frequency $f_{\mathrm{D}} = 66.7\,\mathrm{Hz}$. The index n specifies to the number of selection-combined Rayleigh fading channels. The Rayleigh fading channels are assumed to be independent with identical average received powers, as introduced in Chapter 3. Again, a reversed logarithmic axis is utilized for the presentation of the complementary mission availability approximation $1 - \tilde{M}_n$, highlighting that values on top are superior. The accuracy of the proposed mission availability approximation purely depends on the accuracy of the approximated uptime distribution, presented in Fig. 4.11b. In Fig. 4.15a, the effect of different tolerated downtimes $t_{\max}^{\mathrm{d}}$ for $n = 3$ channels is demonstrated. The curve with no downtime tolerated, $t_{\max}^{\mathrm{d}} = 0$, corresponds to the metric mission reliability, as proposed in [HSF18b]. In this case, interruptions of arbitrary length cause a system failure. Tolerating very short downtimes of $100\,\mathrm{\mu s}$ does not lead to significant improvement. However, by further increasing the duration of accepted downtimes, the mission availability is boosted by several orders of magnitude. For instance, the mission duration of an autonomous car crossing an intersection is about $10\,\mathrm{s}$: The successful accomplishment of this mission can be guaranteed with $97\,\%$ and even $99.999\,998\,\%$ probability, if a maximum downtime of $t_{\max}^{\mathrm{d}} = 100\,\mathrm{\mu s}$

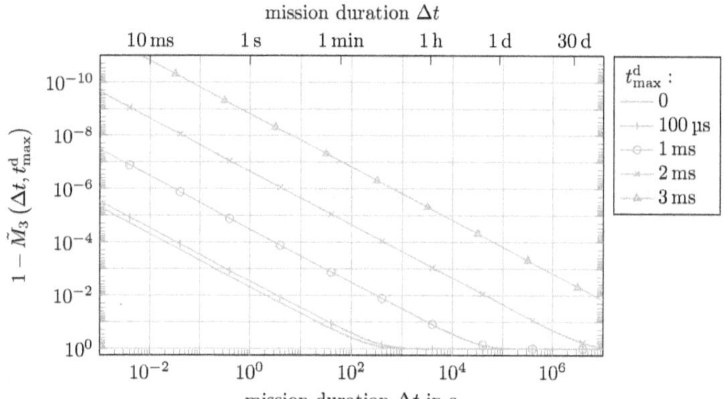

(a) Different tolerated downtimes for $n = 3$ channels.

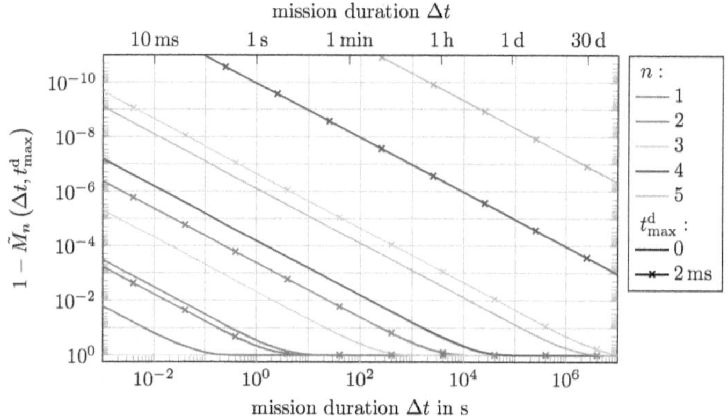

(b) Different numbers of channels n.

Fig. 4.15.: Mission availability for Doppler frequency $f_D = 66.7\,\text{Hz}$ and fading margin $\Phi = 20\,\text{dB}$.

or $t_{\mathrm{max}}^{\mathrm{d}} = 3\,\mathrm{ms}$ is accepted, respectively. Fig. 4.15b illustrates the trade-off between mission duration, number of channels, and tolerated downtime. Here, we concentrate on presenting the proposed mission availability approximation with tolerated downtimes $t_{\mathrm{max}}^{\mathrm{d}} \in \{0, 2\,\mathrm{ms}\}$ for different numbers of channels the user is connected to. For larger numbers of channels, the gain from accepted downtimes increases. It turns out that, e. g., $n = 3$ selection-combined channels which tolerate a short downtime of $t_{\mathrm{max}}^{\mathrm{d}} = 2\,\mathrm{ms}$ outperform the approximated mission availability of $n = 5$ channels with no downtime accepted. Even very long missions of duration $\Delta t = 30\,\mathrm{d}$, which might be of interest for factory automation, can be guaranteed with a probability of $99.999\,98\,\%$ by a diversity system with $n = 5$ channels, tolerating downtimes of $t_{\mathrm{max}}^{\mathrm{d}} = 2\,\mathrm{ms}$. Consequently, the KPI mission availability offers great potential for designing systems with respect to required channels and supported performance guarantees.

4.3 Summary and Conclusions

This chapter has leveraged fundamental dependability theory in order to propel the realization of wireless URLLC by refining the discussion on appropriate KPIs and discussing potential gains due to selection combining. We presented fundamental definitions and derived additional relations between them to put forward a solid foundation for joint discussions and studies on wireless dependability metrics covering aspects regarding uptime, downtime, and mission duration together with success probability:

Concentrating on selection combining of multiple fading channels, is has been revealed that the KPI channel availability and the related popular metric PLR are both of limited benefit for specifying URLLC because they do not describe the influence of the carrier frequency and mobility. Moreover, they fail to reflect the performance with respect to time-related aspects, e. g., time-varying channels or the duration of a certain condition in a wireless system. Consequently, it is of utmost importance to define requirements taking into account the duration of uptime or downtime. For the first time, analytic expressions of the time-based KPIs MUT, MDT, and MTBF have been derived for selection-combined Rayleigh fading channels. Numerical evaluations have demonstrated the influence of the number of channels used for multi-connectivity, fading margin, velocity, and carrier frequency on these dependability metrics and the channel availability. The investigations have been extended assuming Rice fading and interference as causes of failure, discussing the trade-off between fading margin, Rice K-factor, and interference statistics for single- and dual-connectivity, e. g., higher values of the fading margin, Rice K-factor, or number of channels increase the steady-state channel availability, which is bounded due to interference. The value of the optimal fading margin, which minimizes the MDT at a fixed interference outage probability, decreases for larger Rice K-factors. In order to determine concrete performance guarantees, analytic expressions for uptime and downtime distributions have been derived, yielding phase-type distributions and exponential distributions for the considered system model, respectively.

Besides the sojourn times in up and down states, mission duration is identified as a further time parameter to be key for realizing wireless URLLC, characterizing the time interval during which the application does not allow any communication failure. Obviously, the mission duration highly depends on the given application maneuver. This dependability metric has been studied regarding the introduced scenario comprising selection-combined Rayleigh fading channels. Since analyzing the interaction of multiple up- and downtimes is particularly relevant for practical applications, mission reliability has been generalized to the KPI mission availability. This dependability metric specifies the success probability for applications, which can tolerate short communication interruptions. For an exemplary scenario, the trade-offs between mission duration, number of channels for multi-connectivity, and tolerated downtime have been examined with respect to mission reliability and mission availability, respectively. Based on the approach to model the system as a single repairable component, approximations for the mission reliability and mission availability expressions have been developed and their accuracy has been evaluated.

In conclusion, a combined evaluation of average, time-related, and mission-related dependability metrics is of key importance, which has been supported by numerical examples. The benefit of time-related and mission-related KPIs has been elaborated, which are not yet common practice in the wireless communications community, although they enable performance guarantees. Their rigorous utilization has the potential to carry the discussion on requirements of communications systems and their applications to a new stage, because these dependability metrics link the dimensions success probability and individual time intervals. Jointly designing dependable systems from different domains, e. g., wireless communications and factory automation, is a major cornerstone for realizing URLLC in 5G and beyond. In order to successfully introduce wireless communications systems, e. g., to factory automation, it is important to use a common performance evaluation methodology which should comply with dependability theory, as proposed in this chapter. Thus, based on the framework provided in this chapter results can be obtained which help to make conclusions about the dependability of a wireless communications system.

As possible next steps, the discussed dependability metrics can be applied to a system with a packet-based communication standard considering a more realistic channel model including a power delay profile, e. g., as we have demonstrated in [Tra+19b]: In this publication, a simulative dependability analysis of a WLAN is conducted with respect to the proposed metrics in industrial environments, utilizing our empirical channel model [Tra+19a], which accounts for small-scale fading and noise. As a result, the upcoming WLAN standard IEEE 802.11ax is identified as a promising candidate for wireless industrial closed-loop control applications because it provides satisfactory physical layer (PHY) dependability in the considered scenario, especially for applications designed to tolerate at least three consecutive packet losses. For future work, we also propose to use empirical data from communication networks to derive detailed statistical parameters, which, in turn, serve as inputs for the analytical evaluation framework presented. Additionally, latency requirements need to be taken into account as well. The presented analysis can also be transferred directly to other communications systems like 5G to advance the design and the deployment for URLLC, e. g., for industrial wireless closed-loop control.

Dynamic Connectivity for Robust Applications

5

The work presented in this chapter was first published in the paper [Höß+20a]. The author's personal contributions comprise the scenario definition, development of the considered connectivity approaches, implementation and execution of the simulations, selection of KPIs, and evaluation of the resulting data.

The previous chapter demonstrated how multi-connectivity, e. g., selection-combining, can improve wireless dependability by modeling causes of failure as a CTMC. In Section 4.1.6, the term of *robust* applications was introduced, referring to wireless systems with applications that are not sensitive to short outages because many modern applications, such as robotics, can compensate short communication downtimes due to appropriate mechanisms in the application domain [Sch+19a]. As presented in Section 4.2.2, this potential can be leveraged for the design of a wireless communications system, since the importance of a packet for the application depends on the success or failure of immediately preceding packet transmissions. However, existing static diversity schemes are not able to dynamically react to instantaneous failures and therefore waste resources that are not needed for a smooth operation. Thus, this chapter proposes a novel dynamic connectivity concept, namely *reaction diversity*, which reacts to outages due to fading by selecting a different channel. The outage probability of robust applications is studied, which can compensate for short communication interruptions. By means of numerical simulations with Rayleigh fading as the cause of failure, the performance is evaluated in comparison with state-of-the-art approaches such as single-connectivity, multi-connectivity, and channel hopping. The potential of dynamic connectivity approaches (in particular reaction diversity and channel hopping) is demonstrated, which can improve the application outage probability by several orders of magnitude, while utilizing significantly fewer resources than static diversity schemes.

Similar to the system model introduced in Section 3, a set of n channels separated in frequency at least by the coherence bandwidth are assumed, resulting in independent small-scale fading. A single user is assumed to be connected simultaneously to $L \leq n$ channels. Rayleigh-faded channels are considered following Clarke's model [Cla68], whose time correlation properties are characterized by the temporal auto-correlation function

$$\phi_{cc}(\Delta t) = J_0(2\pi f_{\mathrm{D}} \Delta t) \,. \tag{5.1}$$

Here, $J_0(\cdot)$ denotes the zero-order Bessel function of the first kind. The coherence time T_{coh} is the time interval after which the correlation decreases from its maximum 1 to 0.5,

$$|\phi_{cc}(T_{\mathrm{coh}})| = 0.5|\phi_{cc}(0)|. \tag{5.2}$$

Since the time correlation function can be calculated by the inverse Fourier transform of the Doppler power spectrum, there are reciprocal relations between the coherence time and the Doppler spread. In order to estimate the coherence time, the maximum Doppler frequency f_D is often used according to [Rap02]

$$T_{\mathrm{coh}} \approx \frac{9}{16\pi f_D}. \tag{5.3}$$

An outage occurs when the received power level p falls below a threshold $p_{\min}$. In accordance to (4.6), the outage probability of the wireless communications system is denoted as P^{out}. Following from (4.10), the outage probability of one Rayleigh-faded channel is

$$P_1^{\mathrm{out}} = \Pr[p < p_{\min}] = 1 - \exp\left(-\frac{1}{\Phi}\right). \tag{5.4}$$

In order to further characterize the performance of the system, the downtime and uptime distributions are taken into account. Unfortunately, general closed-form solutions of the downtime and uptime distributions for Rayleigh fading are not available. In contrast to the previous chapters, they, thus, will be evaluated by means of simulations.

5.1 Comparison of Connectivity Approaches

As discussed in the previous chapter, diversity is considered to be the key technique to combat small-scale fading. By utilizing space or frequency diversity, the information is simultaneously transmitted on independent fading channels. At the receiver side, a diversity scheme can be used to combine multiple received signals into a single improved signal. However, traditional static diversity schemes utilize many resources, even if the instantaneous channel quality is sufficient, in order to be prepared for degradation on several channels. In addition, static diversity schemes are not designed to react to instantaneous failures, e. g., by providing more resources when conditions are bad. Thus, a novel dynamic connectivity approach referred to as *reaction diversity* is proposed in the following. For comparison, channel hopping and the multi-connectivity scheme selection combining are studied as well. As a baseline, wireless communication over $L = 1$ fixed channel is considered, denoted as single-connectivity.

5.1.1 Multi-Connectivity

When performing multi-connectivity, a user is connected to $L \leq n$ channels and the data is sent redundantly over each channel. In this chapter, the focus is set on the low-complexity scheme selection combining, where the best channel is selected for communication. Assuming that all channels have identical properties, selection combining yields the outage probability

$$P_L^{\text{out}} = 1 - A_L = \left(P_1^{\text{out}}\right)^L = \left(1 - \exp\left(-\frac{1}{\Phi}\right)\right)^L, \tag{5.5}$$

which corresponds to the complementary steady-state channel availability, cf. (4.10). This diversity approach achieves a significantly improved outage probability compared with a non-diversity system with $L = 1$ channel, equivalent to single-connectivity. However, multi-connectivity utilizes many resources, which might not be feasible for scenarios with a high number of users. In the following, the special case of multi-connectivity with $L = 2$ channels is referred to as dual connectivity.

5.1.2 Reaction Diversity

Motivated by the time-correlation properties of faded signals and the fact that robust applications tolerate short outage times, we propose the dynamic, event-driven connectivity approach reaction diversity: A user is connected to $L = 1$ channel. In case of failure, the channel is vacated and communication continues on a different channel. The order of the selected channels is assumed to be pseudo-random and predefined, taking into account that the spectral spacing of successive channels should be large to avoid correlation in frequency. For instance, a wireless closed-loop control application, which is envisioned in the context of URLLC, can then directly send the newest information, instead of re-transmitting the lost and possibly outdated packet. The channel switching procedure is assumed to take place within a reaction time t_{R}. Especially for large coherence times T_{coh} due to low mobility, the effective downtime experienced by the wireless communication system is expected to be reduced by reacting to outages instead of remaining in the faded channel. This procedure therefore also decreases the outage probability P^{out} as well as the application outage probability $P^{\text{A,out}}$ compared with single-connectivity. At the same time, reaction diversity does not utilize more resources than single-connectivity. Of course, reaction diversity needs a feedback channel, which also requires some resources (for a binary information) and can have errors. However, this is out of the scope of this work.

5.1.3 Channel Hopping

Channel hopping is a predefined pseudo-random switching of channels and, thus, it can also be classified as a dynamic connectivity approach [Sal+91]. However, it is not adaptive because channels are continuously changed regardless of their states. We assume the constant sojourn time t_S in each channel. Due to the rapid switching through all n channels, the effective coherence time is reduced to the channel sojourn time t_S. Hence, the effective downtime is also shorter on most cases when compared with single-connectivity. For sufficiently high numbers of channels n, the wireless system experiences outages which are temporally uncorrelated. This is beneficial for robust applications, e. g., control systems which are able to compensate for short outage times. However, on average the resulting outage probability remains unchanged compared to single-connectivity with P_1^out in (5.4) because channel hopping continuously enters and leaves channels regardless of their current states.

5.2 Numerical Evaluation of Connectivity Approaches

In this section, the connectivity approaches reaction diversity, channel hopping, dual connectivity, and single-connectivity are compared based on numerical simulations. The performance metrics outage probability, uptime, downtime, and application outage probability, (as defined in Section 4.1.6) are evaluated with respect to an exemplary medium-mobility scenario comprising velocity $v = 10\,\mathrm{m/s}$ and carrier frequency $f = 2\,\mathrm{GHz}$, yielding $f_\mathrm{D} = 66.7\,\mathrm{Hz}$. In addition, a fading margin of $\Phi = 20\,\mathrm{dB}$ is assumed. All results are obtained from simulated Rayleigh fading sequences with a duration of 30 days at a sampling period $T_\mathrm{S} = 100\,\mathrm{\mu s}$, which is also the transmission time interval (TTI) duration. The filtered Gaussian noise approach was applied to generate the Rayleigh fading sequences [JBS00]. In the following plots, the abbreviations reaction diversity (RD), channel hopping (CH), dual connectivity (DC), and single-connectivity (SC) are utilized.

The outage probability of $P_1^\mathrm{out} = 1\,\%$ is achieved by single-connectivity as well as channel hopping according to (5.4) with the considered fading margin $\Phi = 20\,\mathrm{dB}$ because channel hopping continuously switches channels without accounting for their current states. In this scenario, dual connectivity decreases the outage probability by two orders of magnitude per additional channel resulting from (5.5), i. e., $P_2^\mathrm{out} = 0.01\,\%$. In Fig. 5.1 these values are compared with the outage probability obtained by the proposed reaction diversity approach for different reaction times t_R. It is clearly visible that short reaction times $t_\mathrm{R} < 0.5\,\mathrm{ms}$ significantly reduce the outage probability while still occupying only $L = 1$ channel at a time. The reason is that a channel in outage is swiftly abandoned, which reduces the effective duration of an outage. The outage probability

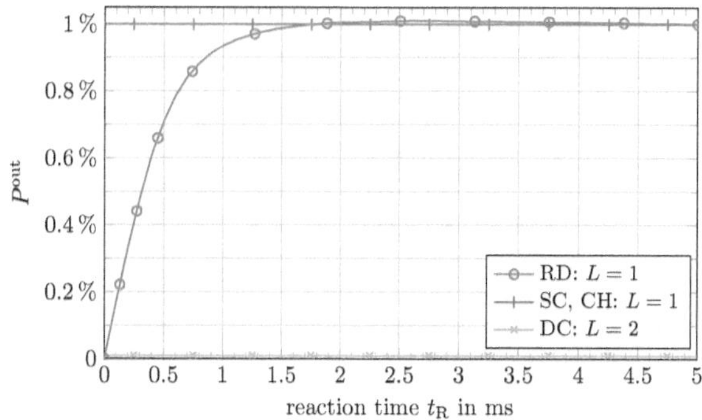

Fig. 5.1.: Outage probability of different connectivity approaches for Doppler frequency $f_{\mathrm{D}} = 66.7\,\mathrm{Hz}$ and fading margin $\Phi = 20\,\mathrm{dB}$.

of reaction diversity approaches the performance of single-connectivity and channel hopping for larger reaction times t_{R}, because it vacates the channel too late, i. e., when it has already recovered from the outage. The slight overshoot reveals that reaction times in the range of the coherence time ($T_{\mathrm{coh}} \approx 2.7\,\mathrm{ms}$ from (5.3)) even lead to a slightly worse outage probability compared with single-connectivity and channel hopping. In these cases the channel usually experiences the complete outage before switching to another channel. However, if we remain on the same channel after the recovery from an outage, the probability of a new outage will be lower than on average due to the shape of the fading autocorrelation. This is corroborated by taking the time-related properties of the fading process into account: The mean downtime (4.26) of a single channel in the considered scenario is $\bar{t}^{\mathrm{d}} \approx 600\,\mathrm{\mu s}$; the mean uptime results in $\bar{t}^{\mathrm{u}} \approx 59.4\,\mathrm{ms}$, confirming the fundamental relation (4.33).

We further delve into the time-related analysis of the fading process and the introduced connectivity approaches by presenting the empirical uptime and downtime CCDFs in Fig. 5.2. The included confidence bounds of $99\,\%$ (dotted lines) are determined by Greenwood's formula [KM58]. The fact that the confidence bounds are tight indicates a low uncertainty of the derived simulation results. Fig. 5.2a illustrates the probability of surpassing a certain uptime achieved by the considered connectivity approaches. Dual connectivity with $L = 2$ redundant channels provides the best performance in terms of uptime. The reaction diversity curves coincide since different reaction times only affect the downtimes but have no impact on uptimes. The uptime CCDFs of reaction diversity are tightly upper-bounded by the one of single-connectivity because reaction diversity utilizes $L = 1$ channel, too. Hence, this uptime analysis immediately indicates how often channels have to be changed applying reaction diversity. The results presented can also be directly transferred to high-mobility scenarios, since the duration values scale inversely proportionally to the maximum Doppler frequency f_{D} and, thus, to the relative velocity v. Channel hopping generates significantly shorter uptimes than the

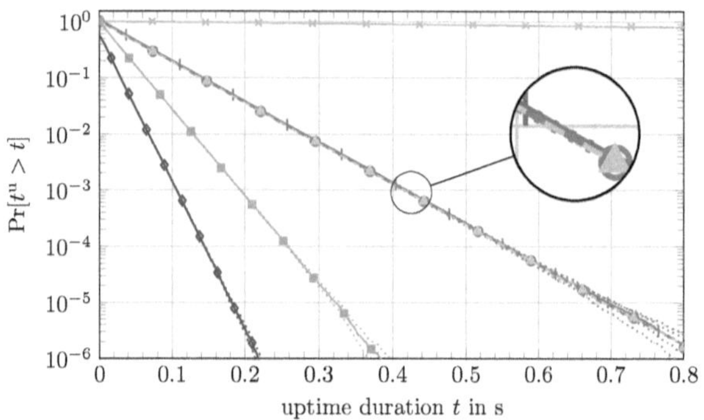

(a) Uptime CCDF.

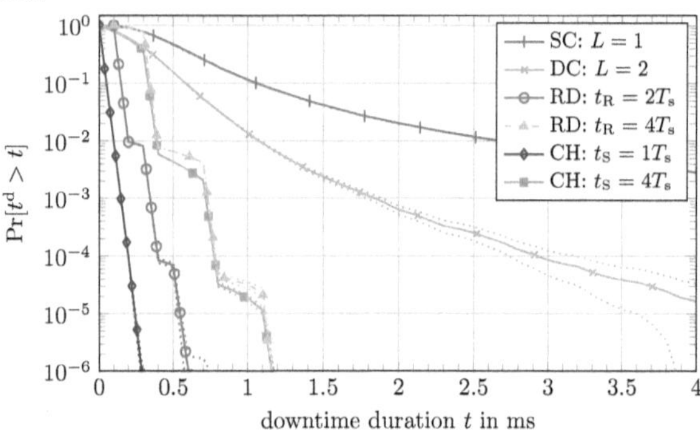

(b) Downtime CCDF.

Fig. 5.2.: Uptime and downtime CCDFs of different connectivity approaches for Doppler frequency $f_D = 66.7\,\mathrm{Hz}$ and fading margin $\Phi = 20\,\mathrm{dB}$. Different reaction times t_R (for reaction diversity) and channel sojourn times (for channel hopping) t_S are presented with respect to the sampling period $T_S = 100\,\mathrm{\mu s}$. At each time instant channel hopping and reaction diversity utilize $L = 1$ out of $n = 4$ channels. The dotted lines indicate the $99\,\%$ confidence bounds. The legend of (b) also applies to (a).

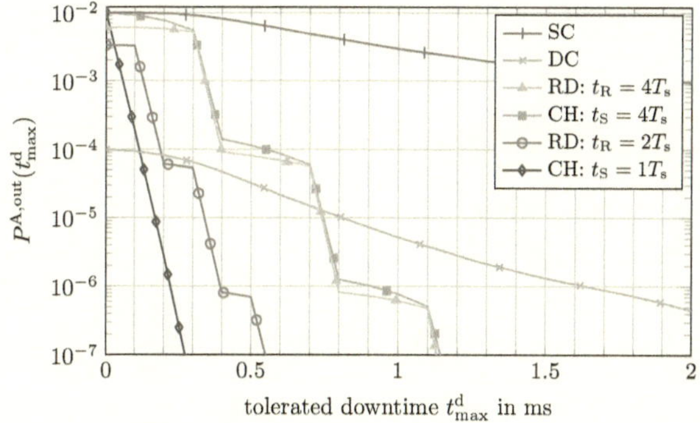

Fig. 5.3.: Application outage probability of different connectivity approaches for Doppler frequency $f_D = 66.7\,\text{Hz}$ and fading margin $\Phi = 20\,\text{dB}$. At each time instant channel hopping and reaction diversity utilize $L = 1$ out of $n = 4$ channels.

proposed reaction diversity approach because channel hopping frequently switches from an operational to a faded channel. Hence, the uptimes of channel hopping decrease for smaller channel sojourn times t_S. As depicted in Fig. 5.2b, single-connectivity shows the highest probability of exceeding a certain downtime due to the auto-correlation properties of the Rayleigh fading process. The gain through dual connectivity with $L = 2$ redundant channels increases with larger downtime. However, both these approaches are outperformed by the dynamic connectivity concepts channel hopping and reaction diversity for downtimes which are larger than the sojourn time t_S or the reaction time t_R, respectively. As expected, switching channels reduces the downtime, which leads to steps in the CCDF. The faster the channel is switched, the lower is the probability of long downtimes. Channel hopping slightly outperforms reaction diversity comparing the same value for reaction time and channel sojourn time, i. e., $t_R = t_S = 4T_s$. The reason is that reaction diversity reacts to any started outage by switching the channel only after the entire reaction time has elapsed. On the other hand, channel hopping continuously switches channels and, thus, leaves a channel in outage on average twice as fast as reaction diversity. A reaction time of $t_R = 2T_s$ is considered to be practically feasible for reaction diversity, because the event of changing channels has to be communicated via the opposite link. Theoretically, the shortest possible reaction time corresponds to one packet length, but the implementation would be very difficult, because each packet would have to be decoded and the acknowledgment would have to be scheduled in the interval between uplink and downlink. On the other hand, channel hopping might be implemented with a channel sojourn time of one packet length because the sequence of channels is predefined.

Studying the downtime duration distributions alone is not sufficient and may even be misleading, because the time duration of outages delivers no statement on the risk of entering an outage. However, the application outage probability, which has been

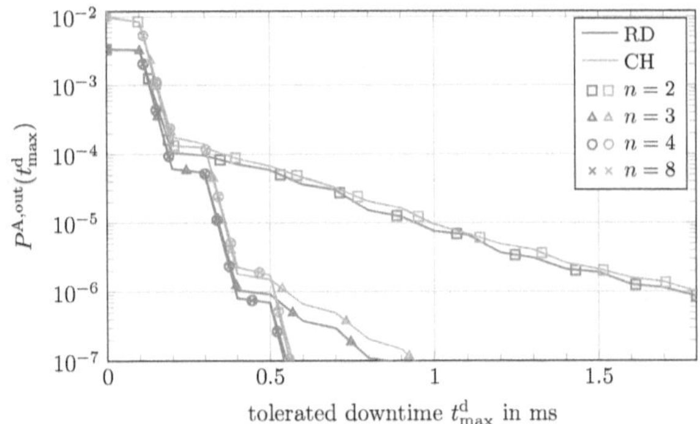

Fig. 5.4.: Application outage probability of different connectivity approaches for Doppler frequency $f_\mathrm{D} = 66.7\,\mathrm{Hz}$, fading margin $\Phi = 20\,\mathrm{dB}$, reaction time/channel sojourn time $t_\mathrm{R} = t_\mathrm{S} = 2T_\mathrm{s}$, and several numbers of available channels n.

introduced in Section 4.1.6, captures this aspect and is presented in Fig. 5.3. In addition, this performance metric provides insights on the impact of the tolerated downtime. Similar to the downtime CCDF, steps are visible for the dynamic connectivity schemes, which stem from the case of subsequently visiting channels which are all in outage. In contrast to the downtime CCDF, the proposed reaction diversity scheme shows gains when compared with channel hopping at $t_\mathrm{R} = t_\mathrm{S} = 4T_\mathrm{s}$, because channel hopping suffers from the potential transition from an operational to a faded channel. The potential of the dynamic connectivity approaches is clearly visible, because every channel switching improves the application outage by two orders of magnitude for the chosen fading margin value $\Phi = 20\,\mathrm{dB}$ because switching to a channel which is in outage only occurs with probability of $P_1^\mathrm{out} = 1\,\%$ in the considered scenario. If downtimes of $t_\mathrm{max}^\mathrm{d} \geq 800\,\mathrm{\mu s}$ can be tolerated by the application, channel hopping and reaction diversity with even $t_\mathrm{R} = t_\mathrm{S} = 4T_\mathrm{s}$ achieve a significantly lower application outage probability than dual connectivity over $L = 2$ channels. At the same time, the dynamic connectivity approaches utilize only half of the resources.

In Fig. 5.4 the effect of different numbers of available channels n on the application outage probability of reaction diversity and channel hopping is demonstrated for $t_\mathrm{R} = t_\mathrm{S} = 2T_\mathrm{s}$. The problem of switching through channels which all are in outage occurs with a probability of $\left(P_1^\mathrm{out}\right)^n$. Hence, if only $n = 2$ channels can be used, immediately successive channel switches cannot significantly improve the application outage probability because of the risk that the original channel is still in outage. In the depicted range, the higher numbers of channels $n = 4$ and $n = 8$ provide the same performance in terms of application outage probability. Hence, a small number of resources can be sufficient to achieve application outage probabilities which satisfy the requirements related to URLLC. This knowledge provides key foundations enabling the design of robust multi-user systems.

5.3 Summary and Conclusion

This chapter demonstrated that dynamic connectivity is a promising concept in order to combat the time-related correlation of failures in wireless communications systems with reduced resource utilization. The novel approach of reaction diversity is proposed, which reacts to outages and, thus, can shorten outage events. By studying the performance metric *application outage probability*, the benefits of KPIs have been demonstrated which reflect the ability of robust applications to tolerate short outage times. Consequently, requirements for other metrics, e. g., packet error ratio (PER), can be relaxed while still achieving the desired performance. These investigations are basic steps to propel the co-design of wireless communications systems and robust applications, which can contribute to addressing key challenges of URLLC applications. Potential topics of future research comprise the evaluation of the proposed dynamic connectivity approaches for other channel models, e. g., simulated Rice fading sequences, as well as coordination issues in the network.

URLLC in Multi-Cellular, Multi-User Systems

6

The work presented in this chapter was first published in the papers [Höß+20b] and [HSF19b]. The author's personal contributions comprise the problem formulation, development and analytical comparison of the considered connectivity approaches, extensions of the stable matching algorithm, implementation and execution of the simulations, and evaluation of the resulting data.

URLLC are considered as one of the key services of the 5G of wireless communications systems and beyond. Enabling URLLC is especially challenging due to the strict requirements in terms of latency and dependability. According to the ITU and 3GPP, URLLC services require an availability[1] of $1 - 10^{-5}$ for delivering a 32-byte packet within $1\,\mathrm{ms}$ [ITU17; 3GP17c]. As discussed in Chapter 4 and 5, multi-connectivity is a powerful approach to increase dependability. However, those chapters and most of the current research focus on single-user scenarios, neglecting the challenges of multi-cellular, multi-user systems, i. e., interference and the competition for limited resources. In contrast to the previous chapters, this chapter does not focus on small-scale fading including time aspects but resource allocation, stability and availability on a higher level of a wireless communications system. This chapter, thus, develops analytic comparisons of different connectivity approaches, showing that multi-connectivity, as defined in this chapter, may not always be optimal in the considered multi-cellular, multi-user scenario. Resource allocation is closely related to matching theory because a matching is an assignment of elements from one set to the elements of another set. In this chapter, novel resource allocation approaches based on stable matching theory are proposed and evaluated to enable wireless URLLC: The pure many-to-one stable matching procedure is extended by utilizing the optimal connectivity approach for each user, optimizing the maximum number of matched resources, and providing a resource reservation mechanism for users suffering from bad channel conditions. System-level simulations confirm that both the proposed matching-based algorithm and its resource reservation extension outperform baseline resource allocation approaches, such as Round Robin, Weakest Selects, and random assignment, in terms of outage probability.

In Section 6.1, existing work related to URLLC in multi-cellular, multi-user systems and stable matching applications in wireless communications are briefly summarized. Sec-

---

[1]The cited references use the term reliability instead. However, their interpretation as a success probability corresponds to the definition of availability in dependability theory, cf. Chapters 2 and 4.

66

tion 6.2 describes the system model, introduces the considered connectivity approaches, and presents the problem formulation. In Section 6.3, the considered connectivity approaches are analytically compared, founded on mathematical proofs. Section 6.4 recapitulates matching theory basics, followed by the proposed resource allocation algorithm and its extensions. System-level simulation results are discussed in Section 6.5, before Section 6.6 concludes this chapter.

6.1 Related Work

This section summarizes research on wireless dependability, focusing on multi-connectivity in multi-cellular, multi-user systems. Then the concept of matching theory and its application within wireless networks are introduced.

6.1.1 Multi-Cellular, Multi-User URLLC

URLLC are considered as one of the key challenges for 5G wireless networks and beyond, receiving major attention from academia and industry. URLLC applications, e. g., wireless factory automation and autonomous driving, combine strict requirements in terms of dependability with latency bounds in the (sub-)millisecond range [Sch+19b]. Recent advances and diverse challenges of URLLC are reviewed in [MMA19; BDP18], examining key enablers and their trade-offs with the conclusion that multi-connectivity, among others, is a promising strategy for realizing URLLC.

However, most contributions are restricted to the special case of a single-user scenario. The few contributions on multi-cellular, multi-user evaluations include the following: The system-level performance of multi-user scheduling in 5G is analyzed in [Ped+18] without emphasis on URLLC. [Poc+18] concludes that fulfilling the URLLC requirements needs novel radio resource management concepts. To the best of our knowledge, there is still a lack of contributions on multi-cellular, multi-user systems which focus on high availability in combination with low latency. In this chapter, a novel analytical framework based on [Sim+19] is developed, showing the potential of matching-theory-based multi-connectivity to achieve URLLC requirements for multiple users.

The following state-of-the-art references on single-connectivity multi-cell multi-user allocation show that the corresponding optimization problems are typically very difficult. Often approximation algorithms are proposed to approach them. In [AP17], online algorithms for the multi-tier multi-cell user association problem that have provable performance guarantees are proposed based on online combinatorial auctions. A two-sided matching market model is utilized in [STH17] to develop an efficient algorithm for user-resource assignments in full-duplex multi-cell networks. The underlying mixed-integer non-linear programming problem is approximated to a geometric problem that is solved by optimality conditions. For multi-cell cooperation in ultra-dense heterogeneous networks, an overview is provided in [KW19].

6.1.2 Matching in Wireless Communications

A situation in which non-divisible goods shall be assigned to entities with different interests can be formulated as a matching problem. One of the most popular matching problems is the stable marriage problem: a set of men and a set of women decide on who to marry based on their preferences over each other, which is a one-to-one matching problem. The notion of stability is important here because it is key for characterizing a robust situation, where no pair of matched partners has an incentive to change the matching. This enables lasting marriages, which are desirable for couples and society at large [GS62]. Stable matchings have first been studied by *Gale* and *Shapley*, showing that there always exists at least one stable matching, which can be constructed by the so-called deferred acceptance algorithm [GS62]. Many-to-one stable matchings have numerous applications, e. g., in the labor market and for college admissions [RS90]. An asymptotic analysis of incentive compatibility and stability in large two-sided matching markets is developed in [KP09].

In wireless communications, resource allocation problems are central challenges due to the limited resources in time, spectrum, and space [HL08]. The first comprehensive tutorial on the use of matching theory for resource management in wireless networks is presented in [Gu+15]. The authors of [Gli16] discuss the application of matching theory for resource management in wireless networks. [Bay+16] provides a comprehensive survey of matching theory, its variants, and their significant properties appropriate for the demands of network engineers and wireless communications. The first application of stable matching in general interference networks is reported in [Jor11]. In HetNets, the assignment of users to their corresponding serving BSs can be modeled as a matching market. In [San+17], the many-to-one stable matching framework is applied to non-orthogonal spectrum assignment with the goal of maximizing the social welfare of the network. A novel rotation matching algorithm is presented in [Di+17] in order to solve the centralized scheduling and resource allocation problem for a cellular V2X broadcasting system with a focus on access latency. Recently, a many-to-many matching algorithm was proposed aiming to guarantee the availability requirements of as many users as possible in a multi-cellular, multi-user system in [Sim+19] , providing a broad overview on wireless multi-connectivity. Resource allocation for URLLC with multiple users based on stable matching is studied in [HSF19b], considering a single cell with small-scale fading. Drawbacks of most of the existing work on stable matching in wireless communications is that either multi-connectivity is not taken into account or only a fixed connectivity approach is utilized. In contrast to previous work available in literature, this chapter focuses on resource sharing in a multi-cellular, multi-user URLLC system in order to obtain a stable matching with the optimal selection among different connectivity approaches. The literature on matching usually applies fixed quotas, denoting the maximum number of matched partners, as the input to matching procedures. The approach taken in this chapter is different in this aspect. Instead of presetting fixed quotas, their values are optimized iteratively, which aims for the simultaneous prevention of underprovisioning and starvation of users. In addition, this work proposes a novel extension to matching-based resource allocation, which specifically covers weak users in order to further increase availability.

6.2 System Model

This section introduces the deployment scenario, defines different connectivity approaches, and presents the optimization problem.

.2.1 Deployment Scenario and Parameters

The focus is set on the downlink transmission of a 2-layer HetNet, where layer 1 is modeled as macrocells and layer 2 as small cells. The HetNet consists of the set $\mathcal{M}$ containing $|\mathcal{M}| = M_c$ hexagonal macrocells overlaid by the set $\mathcal{S}$ of $|\mathcal{S}| = S_c$ small cells. A BS is either a macrocell eNodeB (MeNB) or a small cell eNodeB (SeNB), i. e., $\mathcal{M} \cap \mathcal{S} = \emptyset$. Within the hexagonal macrocellular area, SeNBs are randomly positioned, so that their coverage areas may overlap. It is assumed that MeNBs and SeNBs operate in adjacent sub-6 GHz frequencies, whereby SeNBs operate at the same carrier frequency of bandwidth B. This bandwidth B is equally divided into the set $\mathcal{B}$ of $|\mathcal{B}| = N_B$ subbands. A resource block (RB) is one subband of a single SeNB. This results in a total number of $N = S_c \cdot N_B$ RBs, each of bandwidth $B_{RB} = B/N_B$. All RBs are collected in the set $\mathcal{W}$. The MeNBs a resource-fair scheduler is assumed. A set $\mathfrak{U}$ of $|\mathfrak{U}| = U$ UEs is randomly dropped within the cellular network, whereby a hotspot deployment is considered according to [3GP10].

According to the 3GPP standard, a UE u performs reference signal received power (RSRP) measurements [3GP18]. The MeNB providing the largest RSRP becomes its serving MeNB m_u on condition that a defined required minimum RSRP threshold is achieved. Otherwise, UE u cannot establish a wireless connection, if the largest RSRP is below the threshold. Based on a pre-defined timing structure, the UE u sends the RSRP measurements to inform its serving MeNB m_u about its list of potential BSs $\mathcal{C}_u^{pot}$. This list of potential BSs contains identifiers of BSs in a ranked order according to the RSRP values. It is assumed that the link to MeNB m_u is only used for exchanging control information. Especially, UE u's serving MeNB m_u manages connections of UE u to one or more SeNBs based on the RSRP measurements. Allowing connections to several SeNBs extends the concept of dual-connectivity. For the initialization of links to SeNBs, MeNB m_u sends UE u's access requests to potential SeNBs in the set C_u^{pot}. The set of SeNBs which accept the access request to serve UE u is denoted by $S_u \subseteq S$. The set of all cells serving a UE u results as $C_u = \{m_u\} \cup S_u$. The simulated channels take into account path loss, shadowing, and antenna gains, which rely on the sub-6 GHz channel model and 2×2 multiple-input multiple-output (MIMO) according to [3GP10].

The considered traffic model is the URLLC traffic model with periodic packet arrivals defined in [3GP17b] under system-level simulation assumptions. This chapter considers a fixed number of URLLC traffic UEs with a file size of $F = 200\,\text{byte}$ and a latency budget of $T_{\text{lat}} = 1\,\text{ms}$. In this context, ITU and 3GPP discuss URLLC requirements with respect to purely notional packet sizes between $32\,\text{byte}$ and $200\,\text{byte}$; the higher value is selected because it is stricter. The (user plane) latency is defined as the one-way time it takes to successfully deliver an application layer packet/message from the radio protocol layer ingress point to the radio protocol layer egress point of the radio interface in either uplink or downlink in the network for a given service in unloaded conditions, assuming the UE is in the active state [ITU17; 3GP17c]. This chapter does not concentrate on latency optimization. Instead, the focus is on resource sharing for URLLC in a multi-user, multi-cell scenario, taking the required latency into account as a constraint. Thus, the proposed approaches can be easily transferred to different latency values.

6.2.2 Connectivity Definitions

In the considered scenario multiple RBs from different SeNBs can be assigned to any UE. In addition, the resulting UE data rate depends on whether the individual RBs of a UE are located in the same SB. The BSs are assumed to always transmit at all RBs. The individual connectivity approaches are described in the following. The determined expressions for the achievable throughput are not bounded by a maximum modulation and coding scheme because the bounding affects extremely high throughputs, which are out of scope for URLLC. Thus, the focus is on the Shannon capacity yielding an insightful theoretical bound.

Single-connectivity (SC)

In this case, a UE is connected to only a single SeNB, which is selected based on the measured RSRP. The SINR of UE u, which is connected to *one* SeNB s, is obtained as

$$\gamma_{u,s}^{\text{SC}} = \frac{p_s^{\text{t}} g_{u,s}}{\sum_{s' \in S \setminus \{s\}} p_{s'}^{\text{t}} g_{u,s'} + \sigma^2}, \tag{6.1}$$

with p_s^t being the transmit power of SeNB s. The propagation gain between UE u and SeNB s is given by $g_{u,s}$, and σ^2 is the noise power.

UE u's achievable throughput from SeNB s is computed as

$$r_{u,s}^{\text{SC}} = k_{u,s} B_{\text{RB}} \log_2 \left(1 + \gamma_{u,s}^{\text{SC}} \right), \tag{6.2}$$

with $k_{u,s}$ as the number of RBs assigned to UE u from SeNB s, thus treating interference as Gaussian noise. Each RB has the bandwidth B_{RB}.

Multi-connectivity (MC)

In this case, a UE u establishes links to multiple SeNBs and the assigned RBs are located on different subbands. The set of assigned SeNB is denoted as $\hat{S}_u \subseteq S$. It is assumed that the data/control signals are not transmitted simultaneously from all small cells in $\hat{S}_u$ since the assigned RBs are not located on the same subbands. UE u's throughput is defined by

$$r_{u,\hat{S}_u}^{\text{MC}} = \sum_{s \in \hat{S}_u} r_{u,s}^{\text{SC}}. \tag{6.3}$$

The special case of $|\hat{S}_u| = 1$ reduces MC to SC. It is important to note that this MC definition is a different concept than the multi-connectivity in the other chapters.

Joint transmission (JT)

In this case, a UE u is connected to multiple SeNB and the assigned RBs are located on the same subbands. The set of assigned SeNB is denoted as $\hat{S}_u^b \subseteq S$. It is assumed that the data signals are transmitted simultaneously from all small cells in $\hat{S}_u$ on the same subband $b \in \mathcal{B}$, which has the bandwidth of one RB B_{RB}. This coordination scheme corresponds to Coordinated Multi-Point JT. This connectivity approach is referred to as JT, which results in coherent combining of the received signal, assuming that the received powers add up and the signal components of the SeNBs fall within the cyclic prefix. The corresponding SINR of UE u in subband b is defined as

$$\gamma_{u,b} = \frac{\sum_{s' \in \hat{S}_u^b} p_{s'}^t g_{u,s'}}{\sum_{s' \in S \setminus \hat{S}_u^b} p_{s'}^t g_{u,s'} + \sigma^2}. \tag{6.4}$$

UE u's throughput assigned to the set of SeNBs $\hat{S}_u^b$ over the same subband b is defined by

$$r_{u,b}^{\text{JT}} = B_{\text{RB}} \log_2 \left(1 + \gamma_{u,b} \right). \tag{6.5}$$

If a UE u is assigned to several subbands, UE u's JT throughput aggregates to

$$r_u^{\mathrm{JT}} = \sum_{b \in \mathcal{B}} r_{u,b}^{\mathrm{JT}}. \tag{6.6}$$

Small Cell Connectivity Generalization

UE u's throughput definition (6.6) can be utilized to cover all previously introduced connectivity approaches,

$$r_u = \sum_{b \in \mathcal{B}} B_{\mathrm{RB}} \log_2 \left(1 + \gamma_{u,b}\right). \tag{6.7}$$

Each subband b with no SeNBs assigned to UE u does not contribute to UE u's throughput since $|\hat{\mathcal{S}}_u^b| = 0$ indicates $\gamma_{u,b} = 0$. The set of subbands in which at least one RB is allocated to UE u is referred to as UE u's subbands $\mathcal{B}_u \subseteq \mathcal{B}$. MC is captured if the number of SeNBs assigned to UE u is one for any of UE u's subbands $\mathcal{B}_u$, i.e., $|\hat{\mathcal{S}}_u^b| = 1 \, \forall b \in \mathcal{B}_u$. SC corresponds to the case where a unique SeNB s is assigned to each UE u on all subbands b, i.e., $\hat{\mathcal{S}}_u^b = \{s\} \, \forall b \in \mathcal{B}_u$.

6.2.3 Problem Formulation

In the multi-cellular system considered, multiple UEs aim to satisfy their individual service requirement in terms of throughput by optimizing the number of links and the assigned RBs. This optimization is assumed to be performed by the MeNB. Moreover, the resource allocation should ensure stability, i.e., no matched pairs of UEs and RBs have an incentive to swap partners. Due to the limited number of RBs, a minimal resource consumption is targeted. Thus, the optimization problem is formulated in (6.8), which explains as follows. The objective is to minimize the sum of the numbers of RBs n_u assigned to any UE $u \in \mathcal{U}$, expresssed in (6.8a), such that the resulting matching is stable (condition (6.8b)). The corresponding definitions are presented in Section 6.4.1. In addition, each UE u's throughput r_u (6.7) satisfies the minimum throughput requirement $r_u^{\min} := F/T_{\mathrm{lat}}$ by small cell connectivity (condition (6.8c)). Condition (6.8d) implies that the transmit power p_s^{t} of cell c should not exceed the maximum transmit power $p_s^{\mathrm{t,max}}$. Finally, condition (6.8e) guarantees that UE u's serving cells $\mathcal{S}_u$ are selected from the set of potential BSs:

$$\min \sum_{u \in \mathfrak{U}} n_u \tag{6.8a}$$

subject to:

$$\psi \text{ is a stable matching,} \tag{6.8b}$$

$$r_u \geq r_u^{\min} := F/T_{\text{lat}} \qquad \forall u \in \mathfrak{U}, \tag{6.8c}$$

$$p_s^{\text{t}} \leq p_s^{\text{t,max}} \qquad \forall s \in \mathcal{S}, \tag{6.8d}$$

$$\mathcal{S}_u \subseteq \mathcal{C}_u^{\text{pot}} \qquad \forall u \in \mathfrak{U}. \tag{6.8e}$$

Solving this optimization problem comprises multiple aspects: In the considered multi-user, multi-cellular system, each user throughput depends on the set of serving SeNBs, the number of assigned RBs and the connectivity approach, which specifies the way the RBs are allocated and combined among the frequency subbands. This implies that the optimal connectivity approach for each UE forms the basis for the overall matching outcome of the resource allocation. Thus, different connectivity techniques are compared in Section 6.3 before the proposed stable matching procedures are presented in Section 6.4, which in turn select the optimal connectivity approach for a given scenario realization built on the analytical findings gained in Section 6.3.

6.3 Comparison of Connectivity Approaches

In this section, analytic comparisons between all introduced connectivity approaches are derived. The question to be answered is: "Which connectivity approach should be used from a single user's perspective in order to achieve the highest data rate with a finite number of resources?". Here, it is assumed that the UE's control plane is in the MeNB and the data transmission is performed by the SeNB(s), which are selected according to the RSRP. It is assumed that all SeNBs transmit on all subbands, corresponding to full frequency reuse or full interference.

6.3.1 Single-Connectivity vs. Multi-Connectivity

In this section, the SC and MC for a single user are compared.

Theorem 1. *For any set $\hat{\mathcal{S}} \subseteq \mathcal{S}$ of SeNBs, which establishes MC to UE u, there exists at least one SeNB $s \in \mathcal{S}$, which achieves at least the same throughput by SC allocating the same number $k = |\hat{\mathcal{S}}|$ of RBs, i. e.,*

$$\forall \hat{\mathcal{S}} \subseteq \mathcal{S} \, \exists s \in \mathcal{S} \text{ such that } r_{u,s}^{\text{SC}} \geq r_{u,\hat{\mathcal{S}}_u}^{\text{MC}}. \tag{6.9}$$

Proof. The maximum throughput for UE u in case of SC in one subband is achieved by assigning the SeNB $s \in S$ with the highest received power because of the full interference assumption. The same holds true for all subbands. Thus, assigning the SeNB $s \in S$ with the highest received power for each of the k allocated RBs results in the special case of MC where $S = \{s\}$ which is equivalent to SC via the SeNB s with $k = k_{u,s}$. Consequently, this SC assignment constitutes an upper bound compared to other MC assignments, which completes the proof. $\qquad\square$

This means that a user always has the possibility to waive MC and achieve at least the same performance without the need for more RBs by connecting to only one SeNB. Obviously, this option is preferable in a single user scenario because SC is less cumbersome to realize. However, if multiple UEs are located close to one SeNB, then this selection will lead to congestion.

6.3.2 Multi-Connectivity vs. Joint Transmission

This section presents the performance comparison of MC and JT for a single user. JT is expected to outperform MC since it features coherent combining, which can be confirmed as follows:

Theorem 2. *If one RB of each SeNB in the set $\hat{S}_u = \hat{S}_u^b \subseteq S$, with $|\hat{S}_u| = |\hat{S}_u^b| = \hat{S} > 1$, is allocated to a UE u, JT achieves a higher throughput than MC,*

$$r_{u,b}^{\mathrm{JT}} > r_{u,\hat{S}_u}^{\mathrm{MC}}. \tag{6.10}$$

Proof. See Appendix A. $\qquad\square$

Consequently, if RBs of multiple SeNBs are allocated to a user, they should preferably be arranged in the same subband. In this case, a throughput gain due to JT is achieved because interference is turned into useful signal energy. Of course, there is some overhead in joint signal processing, i.e., data fusion and distribution among SeNBs. Furthermore, JT requires stricter PHY assumptions, e.g., channel knowledge at the transmitter and tight synchronization among the SeNBs.

6.3.3 Single-Connectivity vs. Joint Transmission

In this section, the performance of SC and JT for a single user is compared.

Theorem 3. *If $k > 1$ RBs are allocated to UE u, then SC via SeNB $\check{s} \in \mathcal{S}$ outperforms JT via the set $\hat{\mathcal{S}}_u^b$ containing k SeNBs over one subband b, i. e.,*

$$r_{u,\check{s}}^{\mathrm{SC}} > r_{u,b}^{\mathrm{JT}}, \quad \textit{iff} \tag{6.11}$$

$$p_{\check{s}} > \sqrt[k-1]{\frac{\left(\sum_{\substack{s' \in \mathcal{S} \\ s' \neq \check{s}}} p_{s'} + \sigma^2\right)^k}{\sum_{s' \in \mathcal{S} \setminus \hat{\mathcal{S}}_u^b} p_{s'} + \sigma^2}} - \sum_{\substack{s' \in \mathcal{S} \\ s' \neq \check{s}}} p_{s'} - \sigma^2, \tag{6.12}$$

where UE u's received powers from all SeNBs are denoted as $p_s = p_s^{\mathrm{t}} g_{u,s}$ with $s \in \mathcal{S}$.

Proof. See Appendix B. $\qquad\qquad\square$

Unlike the previous connectivity comparisons, it cannot be concluded here that there is always a SC option which offers a higher throughput than JT. The particular result depends on the combination of the actual deployment, path loss, and shadowing with regard to the different SeNBs. However, SC appears to be more powerful in particular situations where the aggregated bandwidth of one SeNB is more valuable than additional weak links. This applies in the case of a dominant SeNB, as illustrated by the following example.

Assume that only one SeNB $\check{s}$ offers a considerable received power from the perspective of a UE u. The received powers from all other SeNBs are, thus, negligible,

$$p_{\check{s}} > 0, \tag{6.13}$$

$$p_{s'} = 0, \; \forall s' \in \mathcal{S} \setminus \{\check{s}\}. \tag{6.14}$$

Hence, (6.12) reduces to

$$p_{\check{s}} > \sqrt[k-1]{\frac{\sigma^{2k}}{\sum_{s' \in \mathcal{S} \setminus \hat{\mathcal{S}}_u^b} p_{s'} + \sigma^2}} - \sigma^2, \tag{6.15}$$

If the dominant SeNB is not utilized for JT, $p_{\check{s}} \notin \hat{\mathcal{S}}_u^b$, the JT throughput yields zero, $r_{u,b}^{\mathrm{JT}} = 0$, which is trivial because then SC cannot perform worse than JT. The non-trivial sub-case $p_{\check{s}} \in \hat{\mathcal{S}}_u^b$, however, results in

$$\sum_{s' \in \mathcal{S} \setminus \hat{\mathcal{S}}_u^b} p_{s'} = 0 \tag{6.16}$$

due to condition (6.14). Consequently,

$$r_{u,\check{s}}^{\mathrm{SC}} > r_{u,b}^{\mathrm{JT}}, \quad \text{if} \tag{6.17}$$

$$p_{\check{s}} > 0 \tag{6.18}$$

is obtained, which shows that SC outperforms JT in cases with a single dominant SeNB.

On the other hand, JT deserves preference in situations where the received powers of the allocated SeNBs are similar to each other, which is demonstrated for the special case of equal received powers, $p_{s'} = p_{\breve{s}} \; \forall s' \in \mathcal{S}$.

It follows that

$$\sum_{s' \in \mathcal{S} \setminus \hat{\mathcal{S}}_u^b} p_{s'} = (S_c - k)p_{\breve{s}}, \tag{6.19}$$

$$\sum_{\substack{s' \in \mathcal{S} \\ s' \neq \breve{s}}} p_{s'} = (S_c - 1)p_{\breve{s}}, \tag{6.20}$$

because $|\mathcal{S}| = S_c$ and $|\hat{\mathcal{S}}_u^b| = k$. Inverting the relation of (6.11) leads to

$$r_{u,\breve{s}}^{\text{SC}} < r_{u,b}^{\text{JT}}, \quad \text{iff} \tag{6.21}$$

$$\left(p_{\breve{s}} + (S_c - 1)p_{\breve{s}} + \sigma^2\right)^{(k-1)} < \frac{\left((S_c - 1)p_{\breve{s}} + \sigma^2\right)^k}{(S_c - k)p_{\breve{s}} + \sigma^2}. \tag{6.22}$$

The assumption of received powers which are significantly higher than the noise level,

$$p_{\breve{s}} \gg \sigma^2, \tag{6.23}$$

allows simplifications according to

$$(S_c p_{\breve{s}})^{k-1} < \frac{\left((S_c - 1)p_{\breve{s}}\right)^k}{(S_c - k)p_{\breve{s}} + \sigma^2}. \tag{6.24}$$

In the following, two sub-cases have to be distinguished: If all SeNBs are utilized for JT, $k = S_c$ implies

$$(S_c p_{\breve{s}})^{S_c - 1} < \frac{\left((S_c - 1)p_{\breve{s}}\right)^{S_c}}{\sigma^2}. \tag{6.25}$$

After algebraic manipulations, it is obtained that JT outperforms SC in this sub-case of utilizing all SeNBs:

$$r_{u,\breve{s}}^{\text{SC}} < r_{u,b}^{\text{JT}}, \quad \text{iff} \tag{6.26}$$

$$p_{\breve{s}} > \frac{\sigma^2 S^{S_c - 1}}{(S_c - 1)^{S_c}}, \tag{6.27}$$

this condition always holds true for the scenario at hand due to the assumption (6.23). For the other sub-case where JT connections to $k < S_c$ different SeNBs are established, inequality (6.24) is rewritten according to

$$0 < (S_c - 1)^k - S_c{}^k + kS^{k-1} = \sum_{\ell=0}^{k-2} \binom{k}{\ell} S_c{}^\ell (-1)^{k-\ell},$$ (6.28)

where σ^2 is waived due to assumption (6.23). In order to demonstrate that this truncated representation of the binomial formula is greater than zero, it is sufficient to focus on consecutive alternating summands and to show that

$$\binom{k}{\ell} S_c{}^\ell > \binom{k}{\ell - 1} S_c{}^{\ell-1}$$ (6.29)

because the summand with the highest order is always positive and so each positive summand compensates its negative successor. The definition of the binomial coefficient is applied yielding

$$\frac{k!}{\ell!(k - \ell)!} S_c{}^\ell > \frac{k!}{(\ell - 1)!(k - \ell + 1)!} S_c{}^{\ell-1},$$ (6.30)

which simplifies to

$$(k - \ell + 1)S_c = \beta S_c > \ell.$$ (6.31)

This relation is always satisfied for the considered scenario because $S_c > k \geq l$ implies $\beta := (k - \ell + 1) \geq 1$. It can be concluded that JT via any subset of SeNBs outperforms SC, $r_{u,\tilde{s}}^{JT} > r_{u,b}^{SC}$, if the received powers of the allocated SeNBs are equal.

6.4 Proposed Stable Matching Connectivity Algorithm

In this section, an algorithm to achieve a stable matching is proposed, which satisfies the formulated optimization problem (6.8), comprising three components: A variant of the deferred acceptance algorithm is utilized to construct a stable matching. In order to minimize the resource consumption, all the UEs' quotas are optimized, which specify the maximum numbers of allocated RBs per UE. In addition, an extension is proposed which guarantees dedicated RBs to the weak UEs if the pure stable matching algorithm is not able to satisfy all UEs.

6.4.1 Matching for Resource Allocation

The considered scenario is modeled as a many-to-one matching game, comprising the two sets of UEs $\mathfrak{U}$ and indivisible RBs $\mathcal{W}$ as two teams of players with $\mathfrak{U} \cap \mathcal{W} = \emptyset$. RBs can be exclusively assigned to any UE. The UE quota q_u describes how many RBs the UE u can have at most. The problem of assigning the RBs to each UE is a many-to-one matching problem. It is assumed that all UEs and RBs act independently, i.e., the matching game is a distributed game. Each UE and RB has preferences on the RBs and UEs, respectively. The notation $u \succ_w \tilde{u}$ means that player w prefers player u over player $\tilde{u}$. The corresponding preference lists of UEs l_u^{pref} with $u \in \mathfrak{U}$ and RBs l_w^{pref} with $w \in \mathcal{W}$ are obtained based on the SINR values. All players aim for a matching with their most preferred partners. Many-to-one matching games are considered focusing on pairwise stability according to the following definitions:

Definition 12. *A many-to-one matching ψ is a mapping from the set $\mathfrak{U} \cup \mathcal{W}$ into the set of all subsets of $\mathfrak{U} \cup \mathcal{W}$ such that for each $u \in \mathfrak{U}$ and $w \in \mathcal{W}$ the following holds:*

1. *$(w) \subset \mathfrak{U}$ and $\psi(u) \subset \mathcal{W}$;*

2. *$|\psi(u)| \leq q_u$;*

3. *$|\psi(w)| \leq q_w = 1$;*

4. *$w \in \psi(u)$ iff $u \in \psi(w)$,*

with $\psi(u)$ and $\psi(w)$ being the set of player u's and w's partners under the matching ψ, respectively.

Condition 1) describes that players w and u are matched with players out of the set $\mathfrak{U}$ and $\mathcal{W}$, respectively. Conditions 2) and 3) guarantee that the number of matched players is at most the same as the players' quota. Condition 4) states that matched partners are mutually associated, which is naturally given in an UE-RB assignment problem: if an RB w is matched to a UE u than this UE u is also matched to the same RB w

Definition 13. *A blocking pair is the pair of player $u \in \mathfrak{U}$ and player $w \in \mathcal{W}$, who prefer each other over some of their partners in the current matching, i.e., $u \succ_w \tilde{u}$ with $u, \tilde{u} \in \mathfrak{U}$ for some $\tilde{u} \in \psi(w)$ and $w \succ_u \tilde{w}$ with $w, \tilde{w} \in \mathcal{W}$ for some $\tilde{w} \in \psi(u)$, respectively.*

Definition 14. *The matching is pairwise stable, if there are no blocking pairs.*

The definition of stability implies that there is no pair of UE and RB which prefer being matched to each other instead of being matched to their current partner.

Algorithm 1 Many-to-one Stable Matching.

Input: UE quotas q_u, preference lists of all UEs l_u^{pref} with $u \in \mathfrak{U}$ and all RBs l_w^{pref} with $w \in \mathcal{W}$

Proposing and Matching:

Step $t = 0$:

 Initialize the ordered set of UE u's temporarily accepted RBs $\mathcal{A}^t(u) = \emptyset$ for $u \in \mathfrak{U}$.

Step t:

 Proposals:

 Every RB not yet assigned $w \in \mathcal{W} \setminus \bigcup_{u \in \mathfrak{U}} \mathcal{A}^{t-1}(u)$ sends a proposal to its most preferred UE $u \in \mathfrak{U}$ (via its MeNB). This index is cleared from the preference list l_w^{pref} of RB w.

 Decisions:

 Denote RBs which proposed to UE u in step t as $\mathcal{P}^t(u)$. UE u keeps the q_u best ranked RBs from $\mathcal{A}^{t-1} \cup \mathcal{P}^t(u)$ with subject to its preference list l_u^{pref} and updates $\mathcal{A}^t(u)$ accordingly.

Output: Stable matching

6.4.2 Deferred Acceptance Algorithm

Every resource allocation problem has at least one stable matching, which can be constructively determined by the so-called deferred acceptance [GS62]. There is only one stable matching if the preferences of both sets of players are strict and depend on the same metric. The resource-proposing deferred acceptance algorithm yields the stable matching with maximum sum-utility [Jor11]. This algorithm is utilized for the considered resource assignment problem. The corresponding pseudo code is given in Algorithm 1. The matching procedure is assumed to be performed at the MeNB, which receives the required input parameters (preference lists and quotas) from UEs and SeNBs beforehand. The resulting matching is stable according to Definition 14 [RS90]. Of course, this does not guarantee that every single player is satisfied by the matching, e. g., if a RB is matched with its least preferred UE because all others have rejected its proposals. The number of iterations of the stable matching algorithm is bounded by the number U of UEs because no RB proposes to a UE twice. After constructing the stable matching, RBs from several SeNBs which are assigned to the same UE are arranged in the same subbands to take advantage of the JT gain, as derived in Section 6.3.

6.4.3 User Quota Optimization

In conventional matching games, the players' maximum numbers of partners – their quotas – are assumed to be fixed values which are known before the matching procedure. This is plausible for typical matching problems, e. g., college admissions and applications in the labor market. However, in the context of the considered resource assignment problem, the UE quota q_u, $u \in \mathfrak{U}$, is a sensitive hyper-parameter, which should be optimized in order to avoid the following cases: If UE u's quota q_u is too small, the assigned RBs are not sufficient to achieve the required rate, $r_u < r_u^{\min}$. On the other hand, too high quotas may cause over-provisioning of some UEs u, $r_u \gg r_u^{\min}$, if too many resources are assigned. However, this increases the risk of starvation in highly loaded systems, i. e., UEs that have already reached their minimum data rate are assigned further resources instead of unsatisfied UEs.

Thus, the following iterative optimization of the individual UE quotas is proposed: The initial values are $q_u = 1 \, \forall u \in \mathfrak{U}$. After obtaining a stable matching from Algorithm 1, the resulting UE data rates are analyzed. The UE quota is incremented for those UEs whose required minimum rates $r_u^{\min}$ are not achieved by the current stable matching. Then, the stable matching is updated with the improved UE quotas. This procedure is repeated until the stable matching satisfies all UEs' required rates or if the assignment does not change. The latter occurs if all additional potential RBs have rejected UE u's proposal. Then, further increasing UE u's quota q_u has no effect. The amendment of user quota optimization does not weaken the stability because only quotas are changed. Basically, by choosing appropriate quotas, the stable matching outcome is steered towards the desired assignment.

6.4.4 Resource Reservation

Especially in highly loaded systems, it is possible that some UEs remain unsatisfied by the stable matching. Weak UEs are detected as those whose throughput resulting from the stable matching does not meet the required threshold. In these cases, we propose to exclude those weak UEs from the stable matching procedure. Instead, the weak UEs are allowed to allocate their most preferred RBs, choosing the optimal connectivity approach according to the findings in Section 6.3. Subsequently, the stable matching algorithm is performed with respect to the remaining UEs and RBs. The combination of the stable matching and the resource reservation leads to the (temporary) allocation. Due to the reduced number of RBs which are available for stable matching, further UEs may become unsatisfied. In this case, they are considered for resource reservation, as well. This procedure is repeated until all UEs' required rates are satisfied or more than a whole frequency band is requested for resource reservation. In the latter case the temporary allocation result is applied which yields the minimal number of unsatisfied UEs. The UEs who are satisfied by resource reservation meet their requirements and are excluded from the stable matching procedure. They therefore do not compromise the stability of the matching between the remaining UEs and RBs.

6.4.5 Complexity and Convergence

In order to analyze the complexity of the presented algorithms, the big O notation is utilized, which provides an asymptotic upper bound on the corresponding growth rates.

Many-to-one Stable Matching: The preference lists are established based on $U \cdot S_\mathrm{c}$ RSRP measurements between all UEs and SeNBs. During the initialization step ($t = 0$), each UE determines the temporarily accepted RBs, yielding $\mathcal{O}(U)$. The complexity of the subsequent proposal and decision phase ($t > 0$) can be expressed by $\mathcal{O}(N \cdot \max\{q_u\})$. The matching algorithm is guaranteed to terminate in a matching after $t \leq U$ steps because no RB proposes to a UE twice. This results in the complexity $\mathcal{O}(U \cdot \max\{U, N \cdot \max\{q_u\}\}) = \mathcal{O}(UN \cdot \max\{q_u\})$ under the reasonable assumption of less UEs than RBs, $U < N$.

User Quota Optimization: Every UE has the initial user quota of $q_u = 1 \, \forall u \in \mathfrak{U}$. There can be at most $N - U$ iterations in total, each with a single quota increment , since there are only $N - U$ resources available after an initialization of one RB per UE. Thus, optimizing the user quota is carried out in complexity $\mathcal{O}(N - U) = \mathcal{O}(N)$ with $U < N$.

Resource Reservation: In the worst case, each iteration of resource reservation leads to a single new unsatisfied UE due to the reduced number of RBs which are left for stable matching. This corresponds to complexity $\mathcal{O}(U)$.

Finally, the overall complexity, including all iterations of user quota optimization and resource reservation, results by concatenating the three algorithms according to

$$\mathcal{O}\left(UN \cdot \max\{q_u\}\right) \cdot [\mathcal{O}(N) + \mathcal{O}(U)] = \mathcal{O}\left(N^2 U \cdot \max\{q_u\}\right) = \mathcal{O}\left(S_\mathrm{c}^2 N_\mathrm{B}^2 U \cdot \max\{q_u\}\right).$$

$$(6.32)$$

The complexity scales linearly with the number of users and the largest user quota, while the number of SeNBs or the resource granularity show a quadratic impact on the complexity. However, the variables have different orders of magnitude: As mentioned above, less UEs than RBs, $U < N$, are assumed to prevent the trivial case of only one RB per UE. The theoretical bound of $\max\{q_u\} <= N - U$ is not reached in the implementation because the user quota optimization terminates if a stable matching has not changed compared to the previous iteration. Moreover, the number of SeNBs is assumed to be smaller than the number of subbands, $S_\mathrm{c} < N_\mathrm{B}$. Consequently, the number of subbands N_B is identified as a main contributor to the complexity. In order to assess whether the execution of the proposed algorithms is practically feasible due the complexity, it is important to distinguish between mobile and static scenarios. A mobility case would require optimization in real time, preferably during each TTI. On the other hand, a possibly increased computation time can be tolerable for a static scenario, as assumed in this chapter, because the algorithms only need to be performed once during initialization. However, even if real computation capabilities were exceeded, the results would provide an insightful benchmark for practical implementation.

Analyzing the convergence of the presented algorithms is more difficult. As mentioned above, the proposed algorithms are guaranteed to terminate after finite number of iterations. However, convergence is defined as the property that the output gets closer and closer to a specific value as the iterations proceed [Haz02]. This is not achieved here because the output may not approach the resulting allocation more and more closely during execution of the algorithm. Especially, resource reservation for all weak UEs can fail, as mentioned in Section 6.4.4. In that case, the best temporary allocation, which resulted from earlier iterations, is applied. This property does not comply with the strict definition of convergence. Hence, further investigating the convergence properties of the presented resource allocation algorithms and improving them accordingly are potential topics for future research.

6.5 Simulation Results

In this section, the parameters of the system-level simulation scenario are introduced and performance evaluation results for single-user and multi-user settings are presented.

6.5.1 Simulation Scenario

In the following, the proposed solutions are validated in a system-level simulator based on the assumptions and parameters defined in [3GP10]. A HetNet with a macrocell is considered consisting of $S_c/3 \in \{1, 3, 5, 10, 15, 20\}$ small cells per macro sector, uniformly and randomly distributed within the macrocellular environment. Two thirds of U UEs are randomly and uniformly dropped inside an annulus with an inner radius of $10\,\mathrm{m}$ and an outer radius of $40\,\mathrm{m}$ around each SeNB s. The remaining UEs are uniformly distributed within the macrocellular area. Periodic URLLC traffic is considered for each UE. Each UE has a file size of $F = 200\,\mathrm{byte}$ to be downloaded within a latency budget of $1\,\mathrm{ms}$ i.e., the required minimum data rate is $r^{\min} = r_u^{\min} = 1.6\,\mathrm{Mbps}\ \forall u \in \mathfrak{U}$. Our system-level simulation results are determined over $3\,000\,000$ random realizations. Further details about the system-level simulation parameters are provided in Table 6.1. The results are not bounded by a maximum modulation and coding scheme because the bounding affects extremely high throughput values, which are out of scope for URLLC.

Tab. 6.1.: Simulation parameters.

Parameter	Value
Cellular layout	Hexagonal grid, 3 sectors per cell
Number of MeNBs	1
Number of SeNBs per sector	$\{1; 3; 5; 10; 15; 20\}$
Carrier frequency	2 GHz
Bandwidth	20 MHz
Number of subbands	100
Subframe duration	1 ms
Traffic model	Periodic URLLC traffic
Required min. UE rate	1.6 Mbps
Simulated UEs per operating point	3 000 000
RSRP threshold	-114 dBm
Max. MeNB (SeNB) transmit power	46 dBm (30 dBm)
MeNB path loss	$128.1 + 37.6 \log_{10}(d/\text{km})$ dB
SeNB path loss	$140.7 + 36.7 \log_{10}(d/\text{km})$ dB
Thermal noise density	-174 dBm/Hz
Shadowing std.	10 dB
MeNB (SeNB) antenna gain	14 dBi (5 dBi)
Transmission mode	2×2 MIMO
Min. dist. SeNB - SeNB (MeNB)	40 m (75 m)
Min. dist. SeNB - UE	10 m
Number of hotspot UEs	$\lceil 2/3U \rceil$
Hotspot radius	40 m

6.5.2 Single-User Performance

At first, the evaluation focuses on a single-user scenario in order to investigate the impact of the small-cell deployment on the user's SINR and data rate. Fig. 6.1 presents box plots of the three strongest SeNBs with respect to the SINR from each UE's perspective comparing different SeNB densities. The bottom and top of a box are the first and third quartiles, respectively. The band inside the box is the median, outliers are depicted as gray markers. As expected, the SINR to the strongest SeNB decreases for a higher number of SeNBs due to increasing interference. On the other hand, the SINR regarding the second and third strongest SeNBs increase at a higher density of SeNBs. For high numbers of SeNBs, the SINR variances are also reduced for each SeNB rank as well as the aggregated data. The results of the two cases with highest SeNB densities, $S_c/3 = \{15; 20\}$, are almost equivalent to each other. The SINR values of the second and third strongest SeNBs are not higher than 0 dB, which directly results from the SINR definition.

Based on these SINR values, the proposed stable matching algorithm with user quota optimization is performed. The resulting user rates for the considered SeNB densities are depicted as CDFs in Fig. 6.2. Since a single user scenario is considered here, each UE

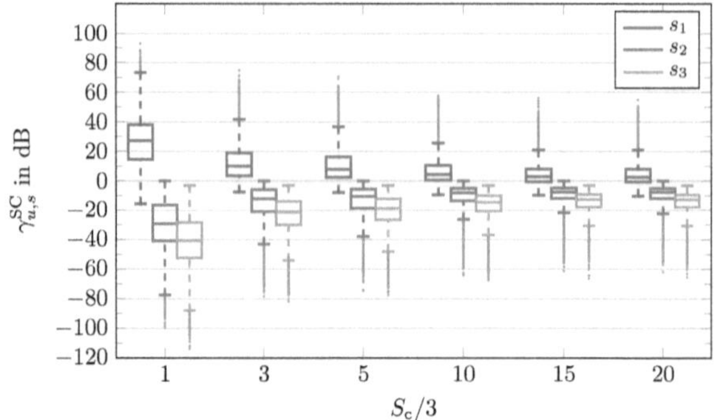

Fig. 6.1.: SINR to the three strongest SeNBs, denoted as s_1, s_2, s_3, for different numbers of SeNBs per sector.

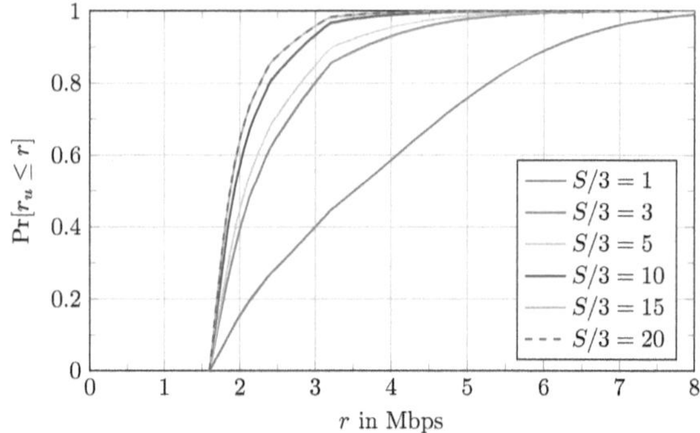

Fig. 6.2.: CDFs of single user rates r_u for different numbers of SeNBs per sector.

is matched to its most preferred SeNB, which is the strongest SeNB in terms of SINR. It can be observed that all rates are higher than 1.6 Mbps, satisfying the required minimum rate. So the maximum availability is achieved in the single user scenario. The quota optimization procedure ensures that no fewer RBs are allocated to a UE than required. At the same time, this aims to prevent overprovisioning of a UE. Since the algorithm stops once the required minimum rate is achieved, some RBs might be left unused. For any percentile, the rate increases for smaller SeNB densities. This is due to the fact that the strongest SeNB provides a higher SINR in less dense deployments, implying higher throughput per RB because the bandwidth of an RB is fixed. The differences between the two cases of highest SeNB density, $S_c/3 = \{15; 20\}$, are not visible, complying with the observation of their SINR values.

The plot in Fig 6.3 shows the average number and the standard deviation of RBs a UE is allocated to, in order to satisfy the required rate. Both values increase for higher SeNB

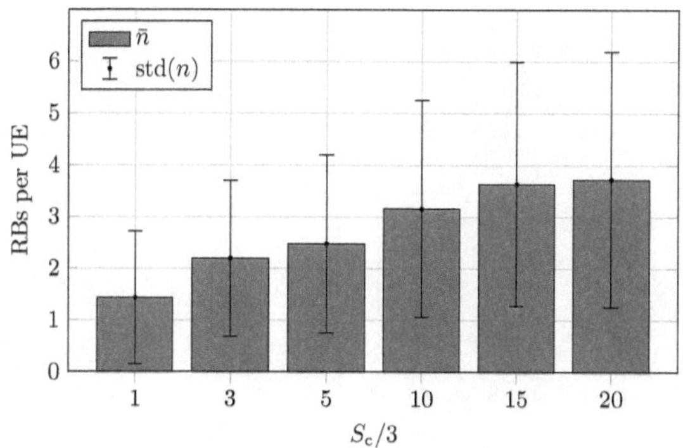

Fig. 6.3.: RBs per UE for different numbers of SeNBs per sector.

Tab. 6.2.: Operating points: numbers of UEs per sector, $U/3$.

$S_c/3$	Medium load (75 %)	Full load (100 %)	Overload (125 %)
1	52	70	87
3	102	137	171
5	151	202	252
10	237	317	396
15	309	413	516
20	402	537	671

densities because of more severe interference. From a single user's perspective, the least dense SeNB deployment scenario seems to be preferable because the required rate can be surpassed the most, utilizing the least resources. However, this is only true if any UE can be served by its most preferred SeNB and if enough RBs are available, such that no competition for resources is fought. Of course, this is fundamentally different in a multi-user scenario, which is more relevant for practical systems.

6.5.3 Multi-User Performance

System-level simulations are evaluated for several multi-user scenarios to demonstrate the performance of the proposed stable matching resource allocation if UEs compete for a limited number of RBs. In order to compare different loads in terms of numbers of UEs, the average numbers of RBs per UE from the single-user scenario are extrapolated: Full load is defined as the number of UEs expected to require all RBs, based on the average over all single user simulations. In addition, medium load (75 %) and overload (125 %) are investigated. The corresponding operating points for all considered numbers of SeNB per sector are summarized in Table 6.2.

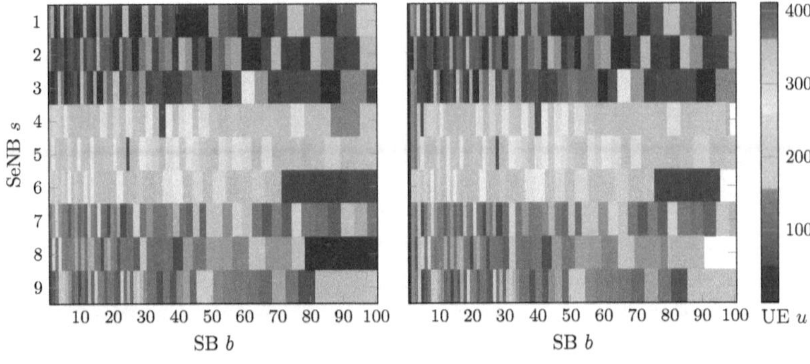

(a) Stable matching.

(b) Stable matching with resource reservation.

Fig. 6.4.: Exemplary resource allocation for three SeNBs per sector at full load achieved by (a) stable matching and (b) stable matching with resource reservation. The color bar in (b) also applies to (a).

In order to demonstrate the performance of the proposed stable matching algorithm and the extension of resource reservation, Fig. 6.4 visualizes exemplary resource allocations for one full load realization with $S_c = 9$ SeNBs. Each row contains the RBs of one SeNB, the columns correspond to SBs. The UE index, a RB is allocated to, is reflected by the color. Without loss of generality, the first/second/third 137 UEs are randomly dropped in sector 1/2/3, which is reflected by shades of blue, green, and red color. Analogously, the SeNBs are distributed to the three sectors. Fig. 6.4a shows the resource allocation resulting from the proposed stable matching algorithm. It can be seen that all RBs are allocated. Due to the quota optimization, the number of allocated RBs differs among the UEs. In general, the number of RBs allocated to a certain UE is low if the UE is matched to an SeNB which offers a high SINR. Allocations of this type dominate the left part of the plot, because the corresponding proposals of RBs are accepted by the UEs as part of the stable matching algorithm. The RBs of a SeNB share the same preference lists because preferences of RBs among the UEs only depend on the SINR value to their SeNB. Due to the competition for limited resources, not every UE can be matched to RBs of the most preferred SeNB. This results in a higher number of requested RBs, equivalent to a higher quota, which is illustrated by wider boxes of the same color. In this particular example, the four UEs assigned to last RBs of SeNBs 3, 6, 7, and 8 are not able to satisfy the required rate, despite of increased quotas. Thus, the proposed resource reservation with respect to those weak UEs is executed. The allocation matrix, obtained after two iterations, is presented in Fig. 6.4b. The guaranteed resources for weak UEs, which are visible in parts of the first columns, ensure that all UEs satisfy the required rate. Thus, some RBs remain unused, depicted as white areas. The resource reservation procedure selects the best connectivity approach, taking advantage of the analytical findings in Section 6.3. In case of UEs assigned to several SeNBs, the corresponding RBs are aligned among the subbands to exploit the JT gain, discussed in Section 6.3.2. Thus, the resource reservation does not utilize complete subbands evenly, e. g., in Fig. 6.4b SeNBs 4, 5, and 6 reserve 5, 3, and 1 RBs, respectively. This effect creates some offsets in the RBs, assigned by stable matching, which are not in the same trend, comparing both resource allocations in Fig. 6.4. Besides that, the allocation is very similar to that obtained by pure stable matching because all UEs except for the weak UEs perform the stable matching algorithm. However, the following tradeoff can be observed: Those RBs which are reserved to the weak UEs are no longer available for the stable matching procedure. Hence, some of the previously satisfied UEs are matched to different SeNBs, which they do not prefer over their initial allocation. They in turn need other resources to meet the minimum required rate r^{min}. For instance, the (green) UE 219 is satisfied by the stable matching with the assignment to SeNBs 1 and 4 on subbands 95 – 100 and 84 – 86, respectively. However, after resource reservation, other UEs occupy subbands 95 – 100 of SeNB 1, who are preferred by the SeNB. Consequently, UE 219 is instead assigned to other RBs from SeNB 4, i. e., on subbands 89 – 97.

In the following, our proposed approaches stable matching (StM) and stable matching with resource reservation (StM+ReR) are benchmarked to three reference resource allocation approaches: the adaptive resource allocation presented in [RC00], which is referred to as Weakest Selects (WeS), Round Robin allocation (RoR), and a random allocation (Ran). All considered resource allocation algorithms are summarized in Table 6.3. In order to extend the performance evaluation of the proposed algorithms, the CDFs of user rates are presented for $S_{\mathrm{c}} = 9$ SeNBs for different loads in Fig. 6.5. The required rate $r^{\mathrm{min}} = 1.6\,\mathrm{Mbps}$ is included to represent the targeted threshold. The included confidence bounds of 99.99 % (dotted lines) are determined by Greenwood's formula [KM58]. The fact that the confidence bounds are very tight indicates a low uncertainty of the obtained simulation results, even in the range of the plotted low percentiles. The probability of not achieving the required rate is denoted as outage probability P_{out}. In the CDF, it corresponds to the percentile of $r = r^{\mathrm{min}}$. The optimal CDF would be a step function from zero to one at the required rate, corresponding to the case that the required rate is met by every UE. In Fig. 6.5b, it can be seen that the rates obtained by the compared algorithms differ significantly. StM outperforms Ran, WeS, and RoR in terms of outage probability. Performing StM+ReR further improves the outage probability by almost three orders of magnitude, resulting in the range of $4 \cdot 10^{-5}$, which is relevant to URLLC use cases. All algorithms except for Ran take the

Tab. 6.3.: Considered resource allocation algorithms.

Abbreviation	Details
StM	Stable matching Algorithm 1 with quota optimization (Section 6.4.3)
StM+ReR	StM with resource reservation (Section 6.4.4)
WeS	Weakest Selects [RC00]
RoR	Round Robin allocation
Ran	Random allocation

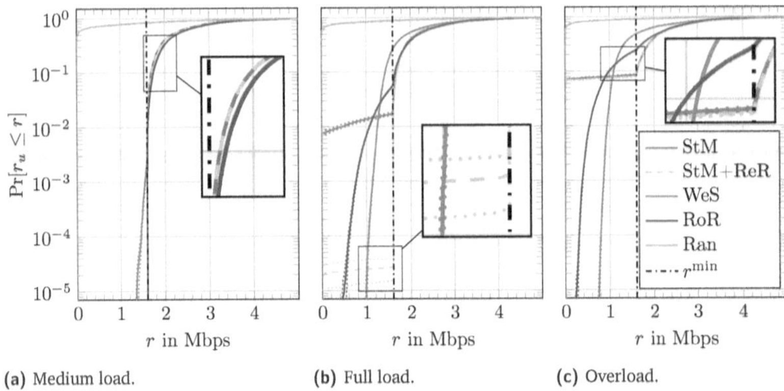

(a) Medium load.   (b) Full load.   (c) Overload.

Fig. 6.5.: CDFs of user rate r_u with three SeNBs per sector at different loads. The dotted lines indicate the $99.99\,\%$ confidence bounds. The legend of (c) also applies to (a) and (b).

targeted rate into account. However, notches in the CDF curves only appear for RoR and the proposed matching algorithms, approaching the ideal result of a step function. The extent of the notches characterizes the sensitivity against the minimum required rate. Any rate below the minimal requirement r^{min} corresponds to an outage, which is reasonable for periodic URLLC traffic. Hence, flat slopes of the CDFs are desirable for $r < r^{\mathrm{min}}$, because it is preferable from the perspective of the multi-user system, it is preferable from the perspective of the multi-user system that a UE unable to reach the target throughput releases its resources to other users. The different slopes below the threshold r^{min} reflect the various ranges of rates for unsatisfied users. The crossing points of multiple curves, however, are not relevant for further performance evaluation in the context of URLLC because they are located below the required minimum rate. Fig. 6.5a shows that for medium load, all resource allocation algorithms improve their outage probability compared to full load, except for Ran. It can be observed that both proposed algorithms, StM and StM+ReR, as well as RoR achieve outage probabilities below 10^{-5}. However, WeS is not able to satisfy all UEs although 25% of RBs are not requested on average in the medium load case. As expected, in overloaded scenarios, all resource allocation algorithms perform worst, as visualized in Fig. 6.5c. The proposed stable-matching-based approaches still outperform all reference algorithms. However, resource reservation can only slightly improve the UE rates due to the lack of resources. In all CDFs, it is visible that the rate of some users greatly exceeds the minimum rate constraint. However, this does not imply that the number of RBs can be further minimized, because the considered RBs are not continuous but discrete resources. Thus, in case of users who do not meet the rate constraint tightly, a single additional allocated RB may lead to exceeding the threshold by far. On the other hand, it is possible, that users with excellent channel conditions greatly exceed the minimum required rate with a single RB.

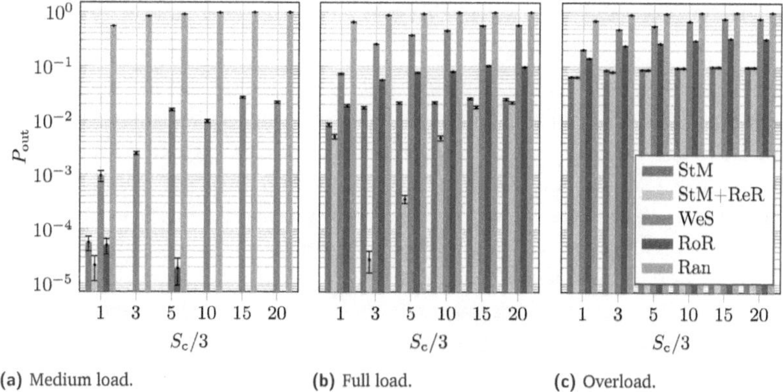

(a) Medium load.　　　(b) Full load.　　　(c) Overload.

Fig. 6.6.: Outage probability for three SeNBs per sector at different loads. The error bars represent the $99.99\,\%$ confidence interval. The legend of (c) also applies to (a) and (b).

After concentrating on one selected deployment scenario, the outage probabilities of all considered resource allocation algorithms are compared for different small cell densities and several user loads as depicted in Fig. 6.6. For full load conditions, Fig. 6.6b shows that the proposed algorithms, StM and StM+ReR, outperform the reference approaches for each considered SeNB density, generalizing the previous observations. The gain through resources reservation differs: Obviously, the overall minimal outage probability is achieved by StM+ReR for three SeNBs per sector. Besides that, the outage probabilities seem to generally deteriorate with respect to higher numbers of SeNB. RoR is the reference algorithms, which performs best, followed by WeS. Regarding medium load, it is clearly visible in Fig. 6.6a that except for the deployment with one SeNB per sector, the proposed resource allocation algorithms achieve outage probabilities lower than 10^{-5}, RoR performs nearly as well. The reason for higher outage probabilities of the proposed stable-matching-based algorithms in the least dense scenario is the fact that the SINR provided by others than the strongest SeNB are weaker compared to the other deployment scenarios. Thus, a UE not matched to RBs of its most preferred SeNB would request a high number of RBs from other SeNB, which cannot be provided in all cases – even at medium load. Fig. 6.6c illustrates the resulting outage probabilities at overload. Again, the proposed algorithms outperform the reference resource allocation strategies. However, the limits of resource reservations are obvious since it hardly improves the outage probability of the pure StM. Similar to the full load case, the reference approaches can be sorted according to their ascending outage probability as follows: RoR, WeS, Ran.

6.6 Summary and Conclusions

In this chapter, a multi-cellular, multi-user heterogeneous system with periodic URLLC traffic is studied, i. e., all users have a stringent availability requirement under a given latency budget. Different connectivity approaches, comprising single-connectivity, multi-connectivity, and joint transmission, are discussed with respect to their availability performance. Analytic comparisons between different small-cell connectivity approaches reveal that multi-connectivity may not be always optimal in the considered scenarios. Thus, insights are provided on how to carefully optimize the resource allocation for wireless URLLC, coping with challenges like interference and competition for limited resources. A novel resource allocation algorithm is introduced, in which a connectivity approach for each user is selected and the maximum number of resources per user is optimized, to satisfy the availability requirements of all users given a latency budget. Hence, the proposed algorithm performs both, link selection as well as sub-band scheduling over single or multiple wireless links. Relying on matching theory, the proposed algorithm is based on stable many-to-one mapping, in which multiple resources are mapped to one user. While stable matching has so far been studied mainly in economics and mathematics, our contributions are the first to leverage it in the context of multi-user, multi-cellular URLLC scenarios with multi-connectivity and shared resources. In contrast to the pure stable matching, which already exists, the proposed approach additionally combines resource allocation with applying the individual optimal connectivity approach for each user and optimizing the maximum number of matched resources, which improves availability. The proposed resource allocation algorithm is extended by a resource reservation mechanism for weak users (suffering from bad channel conditions), which further enhances the overall availability performance. The extensive system-level simulations for different load and density conditions demonstrate that the proposed resource allocation algorithms outperform the baseline resource allocation approaches, such as Round Robin, Weakest Selects, and random assignment, in outage probability by up to three orders of magnitude. Even in a highly loaded system, an outage probability in the range of 10^{-5} is achieved. The proposed algorithms provide enhanced availability satisfying the targeted performance of wireless URLLC.

The work presented in this chapter can be extended by also considering small-scale fading, as we have proposed and evaluated in detail for selection-combined channels in interference-limited environments in [HSF19b]: Then, the preference lists can be determined based on the metric outage capacity. We have shown that an extension of the stable matching procedure, which prevents starvation of weak users, outperforms baseline algorithms regarding the probability of achieving the required minimum rate – as in the case without small-scale fading. Since the findings are similar to those shown in this chapter, only the derivation of the outage capacity expression, which serves as the key input for the stable matching algorithm, is summarized in Appendix C.

In addition, a more realistic scenario could be studied in future work, comprising mobility, different traffic profiles, and resource allocation in uplink and downlink. It would also be of interest to investigate how the proposed algorithms perform in case of frequency-dependent fading if only the average SINR is known. Moreover, several KPIs should be considered jointly, as recommended in Chapter 4.

Conclusions and Outlook | 7

This book provides new insights into understanding and enabling URLLC through multi-connectivity by leveraging dependability theory concepts for wireless communications. Based on the discussed state of the art, relevant causes of failure in wireless communications are modeled, and by deriving and evaluating appropriate KPIs, it is demonstrated that a joint consideration of several dependability metrics is key for the design of future wireless communications systems. Moreover, the main challenges of URLLC in multi-cellular, multi-user wireless systems are addressed, i. e., interference and the competition for limited resources: Different connectivity approaches are compared analytically and novel resource allocation approaches based on stable matching theory are proposed and evaluated. In the following, key results of this book are summarized and conclusions are drawn. Finally, potential topics for future work are recommended.

7.1 Key Results and Conclusions

Dependability theory as a mathematical framework has been providing concepts and methods to analyze and optimize life cycles of technical systems since the 1960s. Dependability theory pillars include accurate terminology, redundancy concepts as well as stochastic models of failures and repairs. Main dependability attributes of technical components and systems are availability and reliability. It is crucial to comprehend their difference: While availability denotes the probability of failure-free operation at a time instant, reliability is defined as the probability of failure-free operation throughout a given time interval. Consequently, referring to reliability without specifying this interval is no valid statement. This book reveals that the common understanding of reliability in the communications sector as the (average) success probability of receiving the correct packet before its deadline corresponds to the definition of availability instead. Although URLLC is targeted in 5G and beyond, research on dependability analyses related to time intervals has hardly been published so far. This book contributes to close this gap and to refine the discussion on dependability KPIs for wireless communications with a focus on URLLC, which will help mastering key challenges of future wireless communications systems, e. g., in wireless factory automation and real-time remote control. This book utilizes dependability theory to foster the understanding and application of relevant terms and concepts for wireless communications by discussing potential dependability gains due to multi-connectivity, which is expected to be a key concept in order to enable URLLC.

It is shown how to model a simplified wireless communications system utilizing selection combining as a system comprising repairable components, represented by a CTMC. Several individual causes of failure and their combination are taken into account, i. e., co-channel interference and small-scale fading based on Rayleigh and Rice fading. Based on the developed CTMCs, this book presents a dependability evaluation and discusses potential improvements due to multi-connectivity addressing average and time-related performance metrics, as well as success probabilities with respect to mission intervals.

Fundamental KPIs are defined and their interrelationships are highlighted to put forward a solid foundation for joint studies on wireless dependability metrics, linking the aspects of uptime, downtime, success probability and mission duration, which are especially relevant for URLLC use cases, e. g., factory automation and robotics. For selection combining of multiple fading channels, it is determined that availability related KPIs, e. g., outage probability and PLR are of limited benefit in the context of URLLC because they do not reflect the influence of the carrier frequency and mobility. Furthermore, they fail to assess the performance with respect to time-related aspects, e. g., time-varying channels or the duration of a certain condition in a wireless system. Consequently, it is crucial to express requirements also in terms of duration of uptime or downtime. This book, thus, derives analytic expressions of the time-based KPIs MUT, MDT, and MTBF for selection combined Rayleigh and Rice fading channels. Evaluations of these dependability metrics and the channel availability demonstrate trade-offs between the of number of channels used for multi-connectivity, fading margin, velocity, carrier frequency, Rician K-factor, and interference statistics. The derived analytic expressions for uptime and downtime distributions, which yield phase-type distributions for the considered system model, enable the formulation of concrete performance guarantees. In addition, the mission duration is shown to be a key time-related parameter in the context of URLLC, because it characterizes the time interval during which an application does not tolerate any communication failure. The mission duration depends on the application and individual sub-tasks or maneuvers and, thus, varies significantly comparing, e. g., industrial automation, robotics, and autonomous driving use cases. The dependability metric mission reliability is proposed and examined for selection-combined Rayleigh fading channels and it is generalized to the KPI mission availability, also addressing applications which tolerate short communication interruptions. Analyzing the interplay of multiple up- and downtimes demonstrates potential dependability gains of several orders of magnitude due to multi-connectivity and the ability to tolerate short communication downtimes. Trade-offs between mission duration, number of channels for multi-connectivity, and tolerated downtime are discussed regarding mission reliability and mission availability, respectively. Moreover, simplifying approximations for the KPIs mission reliability and mission availability expressions are proposed and their accuracy has been validated.

Complementing the concept of multi-connectivity, this book also studies dynamic connectivity approaches which can dynamically change the allocated resources. Dynamic connectivity is shown to be promising in order to cope with the time-related correlation of failures in wireless communications systems with reduced resource utilization compared to multi-connectivity. The novel concept of reaction diversity is proposed, which reacts to fades by changing the channel and can, thus, shorten outage events. This allows relaxed requirements for other metrics, e. g., PLR, while still achieving the desired dependability performance, if short outage times can be tolerated. It can be concluded that a joint consideration of dependability metrics with respect to average success probability, and mission duration as as well as up- and downtime is of major benefit, because it enables performance guarantees. In future, this holistic view of dependability should prevail in wireless communications.

In addition to dependable wireless communication from a single user perspective, this book also considers multi-user, multi-cellular heterogeneous systems with periodic URLLC traffic, i. e., all users have a stringent availability requirement under a given latency budget. This book provides insights on how to carefully optimize the re-source allocation for wireless URLLC in these scenarios, coping with the challenges of interference and the competition for limited resources. Thus, different connectivity approaches, comprising single-connectivity, multi-connectivity, and joint transmission, are discussed with respect to their availability performance. Analytic comparisons reveal that multi-connectivity may not be always optimal in the considered scenarios. The many-to-one stable matching procedure, in which multiple resources are mapped to one user, is extended by utilizing the optimal connectivity approach for each user, optimiz-ing the maximum number of matched resources, and providing a resource reservation mechanism for users suffering from bad channel conditions. Extensive system-level simulations demonstrate that the proposed algorithms outperform baseline resource allocation approaches in outage probability by up to three orders of magnitude. Even in a highly loaded system, availability values in the range of $1 - 10^{-5}$ are achieved. This confirms that the approaches, proposed in this book, provide enhanced availability, sat-isfying the targeted performance of wireless URLLC. Consequently, these contributions can help to enable URLLC, which is expected to constitute a core use case of wireless communications in 5G and beyond.

7.2 Future Work

This work lays foundations for dependable wireless communication through multi-connectivity but does not explore all possible avenues. Among the many different open topics, the following opportunities for further studies are recommended:

- Besides Rayleigh and Rice fading, other small-scale fading models can be taken into account, e. g., Nakagami-m fading, as well as more realistic channel models including different power delay profiles.

- In addition, further causes of failure should be incorporated in the developed model, e. g., shadowing, handover, or coordination issues in the network. Moreover, the uplink has to be investigated, too.

- Empirical data from real communication networks could be used to derive detailed statistical parameters, which, in turn, serve as inputs for the analytical evaluation framework presented.

- The developed continuous-time Markov model can be transferred to the discrete time domain, reflecting packetized communication.

- The discussed dependability metrics can be applied to a real communication standard, e. g., WLAN or 5G, in order to assess the dependability with respect to URLLC scenarios, especially for applications designed to tolerate some consecutive packet losses.

- The investigations on the interplay of up- and downtimes carried out in this book provide impetus to the co-design of wireless communications and control systems (CoCoCo) since control systems are able to tolerate a limited number of consecutive lost packets before ultimately failing. Thus, studies with a focus on KPIs such as mission availability or application availability can contribute to fuse application quality of control and communication quality of service, providing an interface for engineers from both domains.

- Additionally, research on novel resource allocation strategies for URLLC should be conducted, focusing on the proposed time-related dependability metrics. Due to little to no tolerance for retransmissions in URLLC, the development of a real-time multi-connectivity management would be of major interest for 5G and beyond. If it can negatively correlate packet losses in time through dynamic channel allocation, such an approach will yield high dependability, while keeping the average resource consumption low.

- In order to further increase the dependability of wireless communications, potential topics of future research comprise the investigation, enhancement, and demonstration of fading prediction methods. They might enable proactive resource management for URLLC based on the holistic view on dependability of wireless communications presented in this book.